Department of Primary Industries and
Bureau of Resource Sciences

Proceedings

Sustainable Fisheries through Sustaining Fish Habitat

Australian Society for Fish Biology Workshop

Victor Harbor, SA
12–13 August 1992

Editor: D.A. Hancock

Australian Government Publishing Service
Canberra

© Commonwealth of Australia 1993
ISSN 1032-2469
ISBN 0 644 29632 1

This work is copyright. Apart from any use as permitted under the *Copyright Act 1968*, no part may be reproduced by any process without prior written permission from the Australian Government Publishing Service. Requests and inquiries concerning reproduction and rights should be addressed to the Manager, Commonwealth Information Services, Australian Government Publishing Service, GPO Box 84, Canberra ACT 2601.

Preferred way to cite this publication:
Hancock, D.A. (ed.) 1993, Sustainable Fisheries through Sustaining Fish Habitat, Australian Society for Fish Biology Workshop, Victor Harbor, SA, 12–13 August, Bureau of Resource Sciences Proceedings, AGPS, Canberra.

The Bureau of Resource Sciences is a professionally independent Bureau in the Department of Primary Industries and Energy. It was formed in October 1992 from the existing Bureau of Rural Resources and the resource assessment branches of the former Bureau of Mineral Resources.

The Bureau's role is to support the sustainable development of Australia's agricultural, mineral, petroleum, forestry and fisheries industries by providing scientific and technical advice to government, industry and the community.

Bureau of Resource Sciences
PO Box E11
Queen Victoria Terrace
Parkes ACT 2600

Printed for AGPS by Pirie Printers Pty Limited, Fyshwick, A.C.T.

FOREWORD

J.G. Pepperell
President
Australian Society for Fish Biology

This workshop, entitled "Sustainable Fisheries through Sustaining Fish Habitat", continues the Australian Society for Fish Biology's workshop series and its established tradition of bringing together the country's leading experts to freely discuss specific fish and fishery themes of national importance. Past workshops have preceded the Society's Annual Conferences, and began with a meeting on "Australian Threatened Fishes" in Melbourne in 1985. Themes since then have covered diverse topics: "Advances in Aquaculture", "The Use of By-catch Resources in Australia", "Scientific Advice for Managers: Getting the Message Across", "Tagging - Solution or Problem?", "Introduced and Translocated Fishes and their Ecological Effects", "Legal Sizes and their use in Fisheries Management", "The Measurement of Age and Growth in Fish and Shellfish", "Larval Biology", and "Recruitment Processes".

Since 1988, the workshops have been generously supported by the Fishing Industry Research and Development Council (now the Fisheries Research and Development Corporation, FRDC). This support has ensured the attendance of overseas experts at the workshops, and has allowed professional editorial treatment of the published Proceedings. The Bureau of Rural Resources (now, Bureau of Resource Sciences, BRS) has also been most generous in contributing substantially to the costs of publication of Proceedings.

I believe that these workshops, and their published Proceedings, are now recognised as benchmarks in the development of fish and fisheries science in Australia. The present volume is no exception.

The fisheries working group of the Federal Government's Ecologically Sustainable Development strategy noted that ... "management has to establish operating frameworks which embrace both economic and environmental factors ... (which) must ensure prudent management which determines and implements the necessary arrangements and controls to safeguard the resource and the environment in concert with sound development of the commercial industry". (To this I would add the recreational fishing industry). In fact, the first nine of the working group's recommendations relate to concerns for ecosystems and fish habitat.

The primary aims of the Society's successful application for funding to FRDC were: *To identify national fish habitat research needs, and to clarify management's needs to fisheries scientists.* The workshop attracted a record attendance, encouraging lively and enthusiastic debate, and I am confident that all present would agree that those goals were met.

The keynote speaker at the workshop was Stan Moberly, a past President of the American Fisheries Society, and formerly Director of the Alaska Fisheries Department. I am sure you will find his address in these proceedings both stimulating and enlightening, and those present will remember his infectious zeal in relaying his experiences and concerns about fish habitat. In a letter to the Society after his return to the U.S. he stated: "In my opinion, the health of a nation and the quality of life it offers its citizens can be judged by the health of its aquatic ecosystems." I'm sure a large majority of us would agree.

During the workshop, an important new committee on Fish Habitat was formed. Convened by Jenny Burchmore, this committee will play an important role in habitat protection and education in the future.

Rob Lewis swung the resources of the South Australian Department of Fisheries behind the workshop and conference, and his assistance at all levels is greatly appreciated. The workshop organising committee, convened by Barry Bruce, did a magnificent job, so many thanks also to him, together with Ene-Mai Oks, Keith Jones, Mervi Kangas, Gary Jackson and Kate Messner.

I also wish to acknowledge the hard work of the editor, Don Hancock, and his persistence in gently prodding presenters, rapporteurs (and myself) to complete and forward manuscripts so that these Proceedings could be published at all, let alone, on time. He has been ably assisted by Gregg Berry of BRS.

This volume continues the tradition set by previous workshops, and furthers the high standards which they have achieved. I commend it to you.

Contents

Foreword
J.G. Pepperell iii

SESSION 1: Introduction

Keynote Address: Habitat is Where It's At! "It's More Fun to Fight over More Fish than Less Fish"
S.J. Moberly 3

Discussion of Keynote Address
Recorded by J.P. Glaister 14

SESSION 2: A Manager's View of Fish Habitat

Chairperson's Introduction
R.K. Lewis 19

An Ecosystem-based Approach to Marine Fisheries Management
K.S. Edyvane 21

Estuarine Issues from the Manager's Viewpoint
R.H. Winstanley 28

Freshwater Habitat Protection—A Manager's Perspective
P.D. Jackson 32

Discussion of Session 2
Recorded by G.M. Newton 35

SESSION 3: Organisms and Environmental Relationships I—Case Studies

Chairperson's Introduction
P.C. Gehrke 41

Maintain or Modify—Alternative Views of Managing Critical Fisheries Habitat
I.R. Poiner, N.R. Loneragan and C.A. Conacher 43

Rehabilitation, Mitigation and Restoration of Fish Habitat in Regulated Rivers
S. Swales 49

A Habitat Fit for Fish—An Aim of Biomanipulation?
M.T. Bales 57

Enhancement of Estuarine Habitats in Association with Development
R.M. Morton 64

Discussion of Session 3
Recorded by G.A. Thorncraft 70

SESSION 4: Organisms and Environmental Relationships II—Key Variables and Broad Based Issues

Chairperson's Introduction: Key Factors Linking Fish and Their Habitat
B.E. Pierce 75

Freshwater Fish Habitats: Key Factors and Methods to Determine Them
J.D. Koehn 77

Defining Key Factors Relating Marine Fishes and Their Habitats
P.C. Young and J.P. Glaister 84

Defining Key Factors Relating Fish Populations in Estuaries and Their Habitats
N.R. Loneragan 95

Discussion of Session 4
Recorded by M.I. Kangas 106

General Discussion—Day 1
Recorded by D.C. Smith 109

SESSION 5: Impact of Human Activities on Habitat and Fisheries

Chairperson's Introduction
C.M. MacDonald 115

Habitat Changes and Declines of Freshwater Fish in Australia: What is the Evidence and Do We Need More?
M. Mallen-Cooper 118

Ecological Basis for Parallel Declines in Seagrass Habitat and Catches of Commercial Fish in Western Port Bay, Victoria
G.P. Jenkins, G.J. Edgar, H.M.A. May and C. Shaw 124

Effects of Trawling on the Marine Habitat on the North West Shelf of Australia and Implications for Sustainable Fisheries Management
K.J. Sainsbury, R.A. Campbell and A.W. Whitelaw 137

Discussion of Session 5
Recorded by P.C. Coutin 146

SESSION 6: Alternative Uses (Economic, Social and Political)—Key Values

Chairperson's Introduction
R.E. Reichelt 155

Maximising the Potential for Both Sustainable Fisheries and Alternative Uses of Fish Habitat through Marine Harvest Refugia
D.A. Pollard 156

Alternative Uses of Aquatic Habitats: An Economic Perspective
A.J. Staniford 159

Discussion of Session 6
Recorded by J. Robertson 165

SESSION 7: Management (Amelioration, Enhancement, Conservation)

Chairperson's Introduction
B.A. Richardson 171

The Management of Marine Habitat
R.N. O'Boyle 174

Management of the Estuarine Habitat
J. Burchmore 184

Managing Freshwater Fish Habitat
W. Fulton 188

Discussion of Session 7
Recorded by D.A. Pollard 191

GENERAL DISCUSSION

Recorded by P.R. Last 199

SUMMING UP

S.J. Moberly 209

APPENDIXES

Appendix 1: Australia's Threatened Fishes 1992 Listing—Australian Society for Fish Biology
P.D. Jackson, J.D. Koehn and R. Wager 213

Appendix 2: Workshop Program 228

Appendix 3: List of Participants 231

SESSION 1

INTRODUCTION

Keynote address: S.J. Moberly
Rapporteur: J.P. Glaister

KEYNOTE ADDRESS: HABITAT IS WHERE IT'S AT!
"It's more fun to fight over more fish than less fish"

S.J. Moberly

*Northwest Marine Technology Inc.
PO Box 99488
Seattle WA 98199-0488*

Abstract

Coastal resources are under siege. Our fisheries face many threats, and experience at least the following problems; habitat loss, pollution, overfishing, overcapitalization, inadequate funding for management and research, bycatch waste, depleted stocks and increased demand.

The greatest threat to a nation's fishery resource is *not* overharvesting or runaway competition. It is the loss of habitat, and pollution!

Loss of habitat compromises the ability to maintain sustainable long term yields. If a country truly values its fisheries, then habitat must be the centrepiece of the portfolio. Habitat is what provides the benefits. Loss of habitat will eventually doom the fishery aspirations of future generations.

Nations concerned about their fisheries resources, will have in place, national habitat conservation strategies, action agendas, and plans for implementation. Governments must allocate sufficient revenues to assure that habitat conservation objectives can be achieved.

The duty of all fisheries professionals is to take an active, leadership role in the development of "The Fisheries Agenda."

In the United States, fishery leaders, reacting to the serious loss and continuing degradation of habitat essential for fish production, formed a coalition: F.I.S.H., Fishermen Involved in Saving Habitat, a Coalition for the Conservation of Aquatic Habitat. F.I.S.H. has a single mission; to elevate interest in the highest levels of government to the need for a national policy to protect, conserve, enhance, and restore the quality and diversity of fisheries habitat throughout the nation. The American Fisheries Society was responsible for calling the first meeting, bringing the leaders of the nation's fishery community together, which led to the formation of the Coalition.

Our professional society has a special role to play; a leadership role as "honest broker" in advancing fishery science and management. The conservation of aquatic resources requires fishery professionals to be political activists for the resource as well as good scientists.

My duties today are to set the tone of the workshop and to arouse a sense of unity and enthusiasm. What I am about to suggest is an additional role for the fisheries professional. A role most of us are not enthused about. A role some of you downright oppose. I will make a case that we must be proactive for sustainable fisheries and healthy aquatic ecosystems. We cannot sit idly by conducting our research, engaged in lofty intellectual exercises, while habitat disappears. Activism is not for everyone. But our profession must speak out. Our challenge is to maintain our traditional role of providing scientific information while aggressively presenting objective, focused scientific analysis to policy makers.

My definition of fish habitat is fairly simplistic; habitat is anywhere fish are found. We all define habitat in different ways. But to keep it simple, at least for today: habitat is anywhere fish live. The quality and quantity of habitat must be sufficient to sustain healthy aquatic ecosystems.

Our inland and coastal aquatic resources are under siege from careless and unwise land-use activities. Our fisheries face many threats in addition to habitat loss and pollution, including overfishing, inadequate management and research, bycatch problems, and waste.

But, the greatest threat to the resource is not overharvesting or competition among fishermen; it is the loss of habitat, and pollution! Human population growth, ignorance, poverty, irresponsible land use activities and developmental practices have endangered water resources and destroyed habitat essential for sustainable aquatic resources.

Habitat loss compromises and forever destroys our ability to produce sustainable benefits. If any country considers its aquatic resources an asset and values its fisheries, then habitat has to be a priority. Loss of habitat bankrupts the fisheries aspirations of future generations.

All nations, concerned for their fisheries resources, must develop national habitat conservation strategies, action agendas, and plans for implementation. Most important, governments must allocate sufficient *funds* to accomplish the job.

Talk is not action, and saying it does not result in improved habitat! The United States President, during the Earth Summit discussions noted that the U.S. has a set of environmental laws second to none in their stringency. Our federal Coastal Zone Management Act was passed twenty years ago. In the Act, our Congress stated that it shall be a national policy, to preserve, protect, develop with proper environmental safeguards, and where possible, restore and enhance, the resources of the Nation's coastal zone for this and succeeding generations. But, saying it does not make it happen. Our government has not followed through with adequate funding.

While the problems to be solved in the coastal zone are staggering, none are as challenging as habitat conservation. In the U.S. our coastal states and territories responded to the new Coastal Zone Management Act by developing comprehensive management plans. To comply with the provisions of the Act, each coastal state plan was required to include provision for the "protection of natural resources, including wetlands, floodplains, estuaries, beaches, dunes, maritime forests, barrier islands, coral reefs, and fish and wildlife and their habitat, within the coastal zone".

There has been significant progress since the Act was passed twenty years ago but, during the past decade coastal states have witnessed dramatic growth as the American population shifts from the interior to the coastal zone. By the end of this century, seventy per cent of the U.S. population will live within 50 miles of the ocean. These coastal lands are not only our most valuable lands but our most fragile. This population shift will have a destructive impact unless steps are taken to protect valuable coastal resources while accommodating population growth.

Polls over the past 20 years have demonstrated that Americans want increased environmental protection. But these same people participate in one of the industrialized world's most wasteful and polluting life styles. While Americans "talk" a great game, our government hasn't "backed up the talk" with sufficient money to implement the laws that have been passed. Habitat conservation is simply not yet a high priority.

Our budget deficit is talked about daily and it continues to soar out of sight. What is tragic is that the environmental cost of the goods and services we produce is not even accounted for in this deficit. We have a "frontier mentality," which assumes inexhaustible resources and black holes in which to dump our waste. Our leaders during most of the 1980's honestly felt that the American west was still open; "a frontier," and whoever arrived first could claim it. But natural resources are capital; second only to a nation's human resources in value, and a nation's accounting system must include environmental costs. Our Congressmen and Senators are devoted to re-election, and investments now in long term solutions that won't manifest for years does not get you re-elected tomorrow.

We can blame those who destroy our habitat while developing our coast and river basins and we can blame those in Congress that fail to fund legislative solutions, but we can and should also blame the fishermen. Fishermen may be characterized as greedy, wasteful, independent, defiant and firm in their belief that it's not only their privilege to fish, its their God-given *right* to fish! But consider this for a moment. They do act rationally to the rules that govern them. The reason they don't exhibit equity-ownership-behavior is because they don't own anything! We all know that public ownership resources are particularly vulnerable to overspending and this is particularly true with our fisheries.

The United Nations estimated in 1989 that the areas from the coastline to the continental shelf produce 97 per cent of the world's fish. The United States estimates that approximately 75 per cent of the nation's commercial fishery landings (both fish and shellfish) are comprised of estuarine-dependent species. Fishermen are mostly ignorant about the connection between coastal habitats and the fish they pursue.

If fishermen do not understand and appreciate the relationship between the habitat and the fish and shellfish they harvest, don't be surprised. Part of the blame has to fall on us.

Our knowledge of fish and shellfish ecology is substantial and our understanding of the habitat interaction is rapidly increasing, but we have failed to effectively communicate our knowledge to the fishermen and the public in general. Our scientific findings lay dormant in agency reports, peer-reviewed journals, and field notes. We spend most of our time talking to one another. Most of our literature cannot be understood by policy makers. Do we really expect the press or the general public to plough through scientific journals seeking answers to habitat problems?

Fishermen should be in the "front lines," helping to defend against lost habitat. Curbing the loss of habitat will require a change in the behavioral characteristics of fishermen. It will require the scientific community to communicate with those outside our ranks. Politicians will have to do more than pledge concern and then fail to fund appropriate action. Conserving habitat has to be our highest priority. The best and most effective way to communicate what we know is through our professional society. I believe our professional society is the best vehicle to help adjust national and international priorities regarding aquatic habitat.

In the U.S. we have yet another obstacle to overcome; if we were trying to doom our aquatic resources to a long, slow and sure death, we would create a governmental process similar to the way it is now. We have to find ways to work around the inefficiencies of government.

North America is occupied by three countries. Much of the border between Mexico and the United States is defined by a line drawn down the middle of the Rio Grande river. To the north, the U.S. and Canada share the Great Lakes and draw a line down the middle of the St. Lawrence river to define their eastern jurisdiction. In Southeast Alaska the international boundary places watersheds in British Columbia and river mouths in Alaska. Our federal government and the states have drawn an imaginary line three miles offshore to define jurisdictional responsibilities.

As the United States was gradually settled over the last two hundred years, the states selected natural barriers as their boundaries. This is especially true in the Mississippi River basin. Many of the states in the basin claim responsibility to the mid point of a river that defines its border. As a consequence twenty-eight states share responsibility for the resources in the Mississippi River watershed. In most cases the states and the federal government have made strong statements about their commitment to curb the loss of habitat and prevent pollution, but little is accomplished because the institutional arrangements are so varied, so complex, and so illogical.

We use the Mississippi River for just about everything and anything man can imagine to haul up, float down, pump out, and dump in. People who monitor these sort of activities estimate that there is a "dead-zone", approximately a million hectares in size, off the mouth of the Mississippi River. Health statistics reveal that human cancer rates are much higher than the national average for those people living in the lower reaches of the river. Even so, the Mississippi River isn't as polluted as many of the world's rivers.

If our mish-mash of jurisdictions isn't enough to contend with, our federal government divides the responsibility for fisheries and habitat between 37 agencies with 9 executive level departments. It would be difficult to design a more problematic management scheme. But even worse, the federal government as well as many of the states have organized themselves by constituent group (recreational, commercial, aquaculture) or by salinity (freshwater *vs* marine). Fourteen of our coastal states have at least two separate fisheries management agencies and in some cases management of aquaculture is the responsibility of a third agency. In many states the responsibility for water quality and pollution can be found in an additional one or two agencies.

Governments squabble over "states rights" *vs* "federal rights" while what's right for fisheries management is ignored. Who speaks on behalf of the rivers or estuaries? Most of our states don't have an instream flow allocation for fish and wildlife. In dry years it is not unusual to discover that water allocations exceed the total capacity of the river. The scientific community knows that estuaries are users of freshwater. Estuaries require the right amount of water at the right time or they die. But neither fishermen nor policy makers fully appreciate this inescapable scientific fact. It is our duty to educate and motivate them.

In the U.S. the statistics are grim! The most recent National Marine Fisheries Service assessment on the status of our coastal fisheries reveals in part that:

- fourteen stocks, species or species groups are considered to be overexploited;

- nearly 1/3 of all species and stocks for which information is available have experienced population declines since 1977;

- there is insufficient information on the status of another 29% of U.S. fisheries; and

- ten of the 14 overexploited stocks would require 5 to 10 years to recover if fishing stopped all together, yet fishing continues in eight of these ten fisheries.

Despite a stated commitment to habitat conservation, Fishery Management Councils spend most of their time setting catch levels that are more attuned to the economic needs of the fishermen than to long-term conservation of the stocks. The intent of the Magnuson Act, the U.S. federal law which guides management of the nation's living marine resources, will be worthless without habitat protection.

Only recently has the U.S. federal government admitted that overharvest and loss of habitat are the causes for the collapse of many stocks. Also being recognized are state and federal subsidies that stimulate activities that destroy habitat. This is especially true of riparian habitat on federal lands in the western United States where water resources are limited.

In addition to overharvesting, we have a huge problem with bycatch and waste. The shrimp fishery in the Gulf of Mexico takes, in addition to shrimp, 115 species of finfish and the average offshore shrimp vessel takes ten pounds of finfish for each pound of shrimp. This produces a 90 per cent bycatch totalling just under one-half million metric tons (1.1 billion pounds) per year. The bycatch, most of which is dead, is returned to the water. This isn't just an American problem either. According to the United Nations more than 5 million metric tons of sea life were discarded in the world last year; most of it was dead. While fish stocks are being depleted and we are suffering loss of habitat and polluting our waters, our fishing industry continues to promote even greater seafood consumption.

Fishing in the U.S. is the second most popular form of recreation. Our recreational fishing industry continues to promote fishing and the intention is to displace commercial users to meet their demands. The commercial industry expects to meet additional consumer demand with products provided by aquaculture. Our seafood industry launched a promotion last year calling for an increase in domestic consumption of 30 per cent. Their battle cry is 20 by 2000 (20 pounds by the year 2000). A drop in seafood consumption in the U.S. this year is being attributed to consumer concern about the safety of eating seafood. Much of our coastline is polluted.

In two hundred years we have managed to destroy over half of our wetlands, and our fisheries generally have never been in worse condition. But plentiful, high quality functioning habitat doesn't necessarily mean healthy fisheries. The value of a fishery can be driven downward by unthinking political interference and an industry without vision.

Early in my career I decided that the only way I was ever going to see Alaska was to work there. "Fish" was the number one crop in Alaska and fishing was the state's largest private employer.

Habitat and water quality essential to fish in Alaska are virtually intact. Not because the Alaskans are better land managers but because ventures into farming and other agricultural activities by both private business and the government, have mostly been failures, although mining employing old fashioned techniques, and other extractive activities, continue to be a problem. Alaska's population is slightly over 550,000 with half the citizens living in Anchorage. This small number of people hasn't warranted the building of enormous hydroelectric dams. So, many habitat-destroying activities have either failed or just not taken place because of special conditions. But as the state's population increases, unless priorities change, Alaska will no doubt follow the path that other states have taken and find their fish producing capabilities diminished.

When I arrived in Alaska in 1970 I was overwhelmed by the natural beauty, forested mountains and hundreds of miles of streams and rivers! Rivers you could take a drink from. Was I in fisheries biologist heaven?

North American waters are reputed to produce or possess approximately twelve to fifteen per cent of the world's living marine resources. The waters off Alaska contribute about seventy to eighty per cent of this amount which means that Alaska has about ten per cent of the world's living marine resources lying off its shores. What is wonderful and astounding is that virtually all the habitat responsible for this production capability is intact.

In 1973 and 1974 the commercial harvest of salmon in Alaska hit an all time low of about 22 million fish. This was a lot of fish to a biologist from the flat lands of middle America. The harvest had been dropping over the years from a previous high period in the 1930's. Harvest levels exceeded 100 million fish in several years with the peak of 126.4 million fish taken in 1936. These low harvests were attributed to overfishing and less than average environmental conditions.

The low harvest in the early 1970's provided the incentive for a very ambitious rehabilitation and enhancement effort. I started with the program in 1975 as a research biologist and was Director from 1982 until I retired in 1987. The program started in 1971 and was put on a "fast track." By the start of the 1980's the program was in full swing, employing all tactics; clearing streams of logging debris, building fish ladders, and constructing hatcheries of every size. We worked hand-in-hand with a reformed management program that allowed adequate escapement of fish to their natal streams to spawn.

The program was very successful and during the decade of the 1980's, fish production doubled. At the same time Alaska lost approximately 10 per cent of its world market share. Most of this loss was due to farmed salmon which could satisfy the fresh fish markets year round. Salmon production, as a result of the enhancement effort and better management methods, has now pushed the harvest to all time highs. In 1980 the harvest exceeded 100 million fish and the mean harvest for the past twelve years has been 130.5 million. A new historic high harvest was reached last year with 188 million salmon harvested. In the latter part of the 1980's, Alaskan lawmakers passed legislation rejecting salmon farming as a private business. This means Alaska will go head-to-head with the world fresh fish market for about three months and then turn customers away, sending them to other producers that can supply fresh fish on a year round basis. Unless the Alaskan fishing industry finds new markets and/or new product form for its canned and frozen salmon, the benefits to be derived from its enhancement efforts will never be realized.

Having habitat and productive fisheries therefore doesn't mean that all is well socially and economically. But the Alaskans will have time to search for solutions that meet their needs and extract full value from the incredible resource they are blessed with. They have this option only because the habitat hasn't been destroyed.

A second example involving an offshore fishery in the North Pacific and Bering Sea is the Pacific halibut fishery. The native people living along the Pacific coast had been fishing halibut for several thousand years. Commercial fishing was first recorded in 1887 off the coast of Queen Charlotte Islands, British Columbia. By 1915 the halibut stocks off North America were fully exploited. Reduction in catch per effort as the stocks were heavily fished caused demands from the fishermen for the governments to take control of the fishery and assess the condition of the stocks. In 1923, a joint treaty between Canada and the United States established the International Pacific Halibut Commission and over the years the two countries have renegotiated the treaty, gradually giving the Commission more authority to regulate the fishery.

Biologically this resource is managed fairly well. But the Commission controls only seventy per cent of the harvest. Bycatch of halibut in other fisheries, mainly the U.S. groundfish fleet, and waste, account for most of the other 30 per cent which isn't under the Commission's control. When setting the fishing mortality levels, the Commission takes this loss off the top and allocates the remainder to the fishery. Large bycatch coupled with larger than appropriate harvest caused the stocks to decline to an all time low in the mid-1970's. Gradually, a reduction in bycatch combined with lower harvest and favorable environmental conditions allowed the stocks to rebuild to a level which permitted historically high removals in the latter part of the 1980's.

The number of fishermen and fishing vessels participating in the fishery are not under the Commission's control either. In the U.S. the social and economic aspects of the fishery are the responsibility of the North Pacific Management Council, and their actions or inaction have almost destroyed the value of the fishery. The U.S. halibut fishery is managed as an "Olympic event." Open access has allowed a larger than necessary fleet to develop. Since

1975 the fleet has swelled in size from 2,424 vessels to 4,222 vessels in 1990. To maintain harvest limits the fishing season consists of a few 24-hour fishing periods often with individual vessel catch limits imposed. The short seasons cause safety and enforcement problems, and more fishing gear is set that can be recovered. The excess gear with hooked fish is often abandoned, wasting an estimated 1500 mt of halibut. The short seasons do not allow sufficient time to dress the fish promptly, and the product has declined in quality while the cost of participating in the fishery and the processing costs have increased to a level that almost exceed the value of the harvest. The bottom line is that consumers pay a higher price for a product that is lower in quality.

The Canadian fishery, on the other hand, has been closed to entry since 1979 and the fleet has been reduced to 435 vessels. The Canadians adopted individual seasonal vessel quotas in 1991 and the quotas can be marketed in 1993. This will consolidate the fleet further and produce a better product for the consumer.

Fisheries management usually fails because of the lack of adequate stock assessment due to insufficient or poor data and bad science. Couple this shortfall in performance with habitat loss and pollution, and the outcome is predictable. Fisheries management can also fail because of the wrong type of political interference. If government fails to maximize, on a long term basis, the social and economic benefits of the fishery, full value of the resource will never be realized. The ingredients for successful fisheries are good science, good management practices, adequate habitat conservation and the right regulations, together with enlightened political support and the funding necessary to do the job.

The Alaskan salmon fishery and the North Pacific halibut fishery are two examples of situations where the science is good and habitat loss and pollution are not the problem. Full value from these resources is not realized because conventional thought inhibits implementation of new methodologies that would maximize the resource, and government is not providing the leadership necessary to solve the problem.

If your country or mine elects to preserve and manage habitat, it is an investment decision. On the other hand if *government* allows indiscriminate destructive activities it is deciding *not* to make this investment. Open access and Olympic-style fisheries are not the result of random decision making. *Government* is making an implicit decision to overfish and to ignore the waste and social disruption these decisions cause. *Government* is ignoring the dividends that could and should accrue to future generations. The way governments are organized makes managing fisheries, conserving habitat and maintaining water quality extremely difficult. But it was the "rule making process" that created these circumstances and it is only the rule making process that can lead to change for the better. Our fishery resources provide food, enjoyment and employment; they are a litmus test for healthy aquatic ecosystems and this dictates quality of life, the value of which goes far beyond just dollars and cents.

Responsibility for change rests, in large part with you and me. Ultimately the responsibility rests with our elected representatives. We have a responsibility to see that our elected representatives and the public have the opportunity to learn the truth through communication in a form they can understand. This makes us an essential part of the rule making process.

I have no doubt we can do the job if it is a priority. In the U.S., we know how to do good science, we have strong mandates from our citizens, and the politicians have responded by enacting appropriate legislation. But the legislation is not enforced nor is sufficient funding provided to do the job right. What we lack is the will to place enough value on these resources to adjust our priorities, and hold people accountable.

Just last year we watched the U.S. President persuade most of the rest of the world that it was a priority to deal with Iraq. Joined by our allies, we sent half a million of our citizens, half way around the world to fight a war for reasons not fully understood by the majority of our citizens. Government response to a problem doesn't come any stronger. The President said it was the right thing to do and the country followed. Taking care of our aquatic habitat is the right thing to do and if our elected leaders come to believe this, it'll become a national priority. If we can put a man on the moon why can't we bring continuing loss of habitat to a screeching halt?

It is my strong belief that fisheries professionals must play an aggressive role in policy development. We must help influence those decision processes which affect the health and viability of our aquatic resources and which provide for the protection of habitat. Like many in this profession, I didn't come to this conviction naturally or early in my career. I just wanted to study fish. But as the years passed, I couldn't help but notice that fishing wasn't as good as it used to be. Fishermen complained more and consumers were more apprehensive about the safety of eating seafood. Our waters were dirtier and more contaminated and in many places even though fishing was good, you didn't dare eat the fish. When I finally faced up to the fact that I personally could do something about this situation, I became an activist. I used my professional society as my vehicle.

I believe the fisheries professional occupies a special niche; acting as a catalyst and "honest broker", a voice that can speak on behalf of the resource. If we don't speak who will? I now believe we inherit this role when we select this profession; it goes with the territory and its our responsibility. And while some of my colleagues don't believe vacant niches are possible, I contend in this case, that when we fail to occupy this niche, there is no one else to fill the vacancy.

In the early 1980's the American Fisheries Society conducted a member survey and discovered that members favoured more activism by the Society to influence environmental policy and public education. Our members were tired of seeing the resources they worked to conserve continue to diminish. I was an officer in the Society and when we asked who was willing to help, few stepped forward. Activism can be scary! What our members were suggesting was that the Society should be the activist and not the individual members. So we did what any good society of scientists would do and we conducted another survey. The second survey confirmed the first, so we conducted a third survey; again with the same outcome.

So we had our mandate from the members and we set off to be activists. But we were not very sure what needed to be activated. So again, like any scientific society worth its salt, we set about to study the matter. We decided to "study" the federal government. After all, we were upset with the fragmentation of fisheries research and management among federal agencies, and the severe reductions in funding, annually proposed in the President's budget for the National Marine Fisheries Service.

The Society's President appointed a 20-member committee of senior fishery scientists and managers having extensive and varied experience in federal, regional, and state fishery affairs. The President charged the Committee to review the current distribution of major fishery research and management authority among federal agencies; the rationale for the current pattern of organization and services; and to explore various alternatives for improving federal organization and operation for fishery conservation and management. Finally we were to formulate a proposal for reorganization of the U.S. federal government's role in the nation's fishery affairs. We called the committee the Federal Fisheries Responsibility Committee.

I joined the Committee because I was passing through the officers' chairs and if the Committee completed its work during my term as Society President, it would become my responsibility for implementing the recommendations. My foresight proved correct. The Committee completed its work in 1987 and the Society Executive Committee charged me, as incoming President, to implement the Committee's recommendations.

I couldn't very well order another survey nor appoint another committee and I had my orders. I just didn't know how to go about getting the job done. In the past, when difficult and contentious matters faced the Society it was a common and accepted tactic to study the matter by appointing another committee. At least it would come due on someone else's shift. But my choices were limited and besides I was impatient! We had just completed a two-year extensive analysis of fishery problems by an unusually experienced professional group. The study was the best available comparative assessment of fishery-related roles of federal government agencies. The study offered specific recommendations for improving the integration and effectiveness of federal fishery management authority and responsibility. This was good stuff. I had the support of the Society's governing body and I was bound and determined to do something. I just didn't know what. But, to make things happen, you have to start somewhere.

I procrastinated for several months and finally it dawned on me. If we expected to see real change in how our government managed our fishery, we were going to need all the help we could find. I started contacting all the other organizations in the fisheries community and we compared notes. I called all the national conservation and environmental organizations; the commercial and recreational fishing organizations; the media; and several individuals that I knew were interested in better fisheries management. When I felt there was sufficient support to assemble a "critical mass" of these groups and individuals, I called them together in the first such meeting in August 1988. Everyone I invited attended! It seems that all our allies in this fish business were not only concerned but they were willing to help do something about it. We had all been fighting the same fight but had failed to look upon each other as allies. In fact, many of the organizations that were sitting about the table that August were more accustomed to competing with each other for harvest quota.

We quickly reached agreement on one issue. We concluded it was more fun to fight over more fish than less fish. Secondly, everyone agreed that habitat was the single most important issue. We agreed that a nation-wide aquatic education program was the only long-term solution if we intended to change the attitude of an entire nation. There was no consensus on any other issue. The moment a species of fish was mentioned or a type of gear or a geographical location, everyone got into small camps and were either intensely interested or not interested at all. We agreed to work on those issues for which there was a unanimous consensus.

The Society was the catalyst for this first gathering of the **F.I.S.H.** coalition (**F**ishermen **I**nvolved in **S**aving **H**abitat). F.I.S.H. would be a nationwide coalition consisting of groups and individuals representing the interests and concerns of sport fishing, commercial fishing, conservation organizations, fishery scientists, fishery managers, fish processors and distributors and others. The charter members adopted a single mission; to elevate to the highest levels of government conscience the need for a national policy to protect, conserve, enhance and restore the quality and diversity of fisheries habitat throughout the nation. We intended to see that this goal became a priority with our government. There were no illusions about how long this would take. It would be a forever job. Constant vigilance would be necessary to prevent what success we achieved over time from being eroded.

Our goal for aquatic education was to teach our children *how* to think about conservation, not *what* to think. We envisioned a national aquatic education program that would provide every kid in America with knowledge to learn their "environmental address." We wanted to educate fishermen about the connection between habitat and harvest, so as to enlist them in our front line of defence. We wanted urban/suburban planners to become informed of the value of aquatic habitat so they could include maintenance and restoration provisions in their regulatory and permitting processes.

Our venture into the world of fish activism was launched but we were determined to do more. A broader role for the Society required greater visibility to the public. The only power the President of the Society has, is the power to appoint members to committees and write letters and call on important people on the Society's behalf. Using this "power" I challenged the Fisheries Administrators Section to host an international workshop, inviting all the fisheries leaders in North America, to create a vision for our fishing future. Who better to start the ball rolling than the very people that are paid to spend most every day thinking about fish?

The Administrator's Section agreed to host the meeting. Next we persuaded the U.S. Fish and Wildlife Service and the National Marine Fisheries Service to co-host the workshop. This provided the fiscal resources necessary for a successful meeting. Next we invited the National Fish and Wildlife Foundation (NFWF) to join the effort. We suspected that the Foundation might be another "honest broker" and we could use a partner to assist in our efforts. For a young organization they had great credibility and influence with our Congress.

The NFWF is a private, nonprofit organization established by the U.S. Congress in 1984. NFWF encourages and administers private sector contributions in support of the programs of the U.S. Fish and Wildlife Service, and promotes other innovative public and private partnerships to enhance the conservation and management of the nation's fish, wildlife and plant resources. Its annual budget is around $20 million, with $5 million allocated by Congress and the remainder raised from private donations. The NFWF had been conducting "assessments" of the federal agencies most responsible for the nation's fish and wildlife resources. These "assessments" were presented by NFWF to Congress during budget deliberations. Through their efforts the budgets of these agencies were increasing each year and Congress was starting to realize this was a good investment.

The North American Fisheries Leadership Workshop was held at Snowbird, Utah, in May 1991. The workshop product is our vision statement for the future. We have shared this vision with every organization that might have an interest in the nation's aquatic resources. We have requested their comments and we intend to incorporate them into a North American Fisheries Action Agenda. Following the workshop in Utah, we commissioned Dr. John Harville, who had chaired the Federal Fisheries Responsibilities Committee, to provide an analysis of all the major reviews of our national fisheries programs conducted in the past fifteen years. This analysis documented many areas where there was a unanimity of consensus and where immediate action could be taken to improve our fisheries. This too has been widely distributed.

In 1991 we agreed to create a Fisheries Action Network (FAN). This was our next step. Our purpose would be to support and enhance efforts for informed fisheries conservation, restoration and sustainable use. We would develop an interactive process for professionals and user groups to assist in the formulation of federal fisheries program direction. FAN is a (North American) fisheries information communication and advocacy process to inform and motivate conservation groups, the fishing industry and individuals. We reasoned that by providing scientifically sound fisheries information we can help others accomplish their own fisheries programs and goals. As part of our FAN effort we have developed a review process with the National Fish and Wildlife Foundation staff, providing peer review for the 1993 fisheries program proposals advanced by the National Marine Fisheries Service, U.S. Fish and Wildlife Service, Bureau of Land Management, and the U.S. Forest Service. We have prepared congressional testimony for the 1993 budget of these agencies. This past March the Society's Executive Committee established a Legislative Committee to further direct the Society's input into the federal fisheries process.

Since colonial times the fish-producing capability of the U.S. has been significantly reduced. Watersheds have been altered and habitat lost through reasons familiar to us all. Over the 122-year existence of the American Fisheries Society our profession has evolved from fish culturists to fish biologists to fish scientists and we now recognize and embrace a multi-disciplinary approach. And in the past few years we have learned to be activists. The number of fisheries scientists in North America is at an all time high. At the same time, fish stocks have never been in worse condition, half of our wetlands are gone and habitat continues to be unacceptably altered and destroyed. Objectives for the nation's fisheries and aquatic ecosystems remain unfocused and largely unimplemented. Prevention of habitat loss is still not a priority with our government. Our numerous agencies are often at odds and we, as fisheries professionals, have been guilty of not communicating our knowledge to those most important to the decision-making process.

The American Fisheries Society has recognized that there is a critical need to communicate scientifically-based fisheries conservation information to the public. The only basis for a rational fishery is good science. The very nature of our job places us in a special role; a special niche as the "keepers of the data base." The data we collect must be turned into useful information capable of being absorbed and acted upon by our decision makers. While the political process should not be allowed to shape our scientific conclusions, the benefit of our analysis must be reduced to a digestible form for otherwise we will be ignored.

Some professionals fear that we may compromise our reputation or our ability to do good science if we aggressively speak out. I don't agree but, is there an alternative? Our challenge is to maintain our traditional role of providing scientific information while presenting focused scientific analysis to policy makers. We must call attention to the value of these resources; must work to change the priorities and help our government to lead. Most importantly our children must learn their "environmental address," for regardless of our action or inaction, our children will inherit our accomplishments.

DISCUSSION OF KEYNOTE ADDRESS

Recorded by J.P. Glaister
Fisheries Division, QDPI
PO Box 46
Brisbane QLD 4001

Mick Olsen wanted to know what percentage of the budget is ever put towards educational use amongst the fishermen, for example in Alaska? Did Stan Moberly have to use very much of funds allocated to the programme he directed to make the fishermen understand what they were trying to do in fisheries management?

Stan Moberly thought the answer was no. The mandates are given that direct managers to outreach and inform fishermen but when you look at the budgets of the programmes you find that almost all of it is allocated to *collecting data*, not to turning the data into useful information and then outreaching the information to fishermen. Outreach is almost always stated as a programme goal and something managers say they want to do. Stan Moberly sits on the Advisory Committee for the Secretary of Commerce which has responsibility for management of the nation's living marine resources. When you look at the federal program you find strong mandates for protecting and conserving habitat and for outreaching to the fishermen and the public. But when you examine the budget and see how the agency allocates its resources then you see the true picture. What the agency says it is going to do is one thing but what they really do is revealed in the budget process, which is that the agency has allocated little to protecting and conserving habitat and outreach to the public and the fishing industry.

Another thing that complicates the situation further is that fishery biologists do not like doing this work, so they often transfer the task of outreach to an Information Education Section of information specialists. These specialists in turn have a fight to get the biologist to even talk to them - because they won't be busied with them. And often, even in the Information Education Sections, when budgets are shrinking so does the outreach program. They continue to produce the information and do even less outreach. Through the F.I.S.H. ("Fishermen Involved In Saving Habitat") Coalition, on the other hand there is no attempt to create any new information or educational material. There is already sufficient information available if only it can reach people so that they are aware of conditions or circumstances. If people who are concerned about aquatic resources are informed they usually support a higher priority to conservation and protection of these resources.

Peter Young commented that at the International Fisheries Congress in Athens earlier on in the year there was a lot of talk from social scientists, particularly from the USA, complaining bitterly that they had been effectively excluded from the whole fisheries management process. Did Stan Moberly think that in the USA there is an increasing awareness that these people can actually be allies in the whole process and become a very powerful lobby on behalf of these kinds of issues?

Stan Moberly believed they are an essential part of the process. Fisheries are a common resource and belong to the public. In the mid 1980's the American Fisheries Society created its first dual-disciplined section called the bioengineering Section. You wouldn't guess where most resistance in the Society came from - the fish culturists. Fish culturists require expertise and assistance of engineers to construct successful culture facilities. Stan Moberly believed that fish culturists would "pour their own raceways" if they could secure the building permits. Administrative procedures and building codes require that licensed engineers be part of the process of designing and building fish culture facilities. Despite the resistance, the next attempt at a dual-disciplined section was to bring in the anthropologists and the economists to try to form a Socio-economics Section. This effort was successful and occurred without much resistance. The Society was changing. The theme of the Society's 1988 annual conference, held in Toronto, was centred around "what is biologically possible, economically feasible and politically doable," that fish management is a "three-legged" stool. This is the part of the whole system that fisheries scientists haven't been trained to do. We look and observe and collect data and enjoy our jobs but never expect that we would have to transform what we see and know into information beyond our own disciplines. But in modern fisheries management it is absolutely essential that we do this. We have to understand and interact with disciplines such as engineering, economics and sociology. We don't number enough! There are approximately 18,000 fisheries professionals in North America but there are tens of thousands of engineers—and they don't spend very much of their time designing and helping to build fish hatcheries; they're pushing dirt around the shoreline and building condominiums and highways. We need a multi-disciplined approach in managing fisheries as there are too many competing interests for habitat essential for fish. He noted that we are seeing more and more of a multi-disciplined approach in North America and he hoped that it will continue.

Bryan Pierce's opinion was that the education component within South Australia and Australia all over is generally lacking. The Australian Fisheries Service has tried quite a few different approaches. Is there any quantitative information as to which is the best approach towards activating the community response?

Stan Moberly thought there wasn't enough experience for that. He could only point to what had actually worked. It was less a question of obtaining new information, which would have been the easier option, than of circulating and doing outreach with already available material. The difficult part is transferring information to those who need the information so that they can involve themselves in the decision processes. The most powerful process the F.I.S.H. Coalition offers is for getting the diverse, warring components of the fishing community talking to one another and all marching in the same direction at least on one issue—conservation of aquatic habitat. In North America, those in the fisheries community have a reputation of cancelling each other out—sports fishermen lobby their point of view followed by the commercial fishermen followed by the scientists followed by the land developer and so on, by which time the politicians are pretty comfortable in not doing anything. If we expect change and if we expect the political process to work for us we must reach consensus as much as possible and then represent our position in the political process. The members of the F.I.S.H. Coalition reached consensus on the necessity to conserve and protect aquatic habitat and then, when they all marched together with this message, the politicians started to pay attention. When the Coalition lobbied in Washington, the team consisted of a representative from the top recreational fishing groups, one from the National Wildlife Federation (representing the environmental community), one from the commercial fishing industry and one representing the scientists and managers, and the four of them visited Congressmen and Senators. The essential approach is to be seen to be in agreement to lobby for those issues on which consensus can be reached.

Stan Moberly suggested that the Society has two roles—one is its traditional role of scientists and managers, and the other is to function as an "honest broker" to run a neutral podium, to call the meetings, to be the gelling agent, the catalyst and to assist the fisheries community in reaching agreement as much as possible and to work together on those issues they agree upon. He thought that probably this approach has to be the formula, at least this seemed to be working for them in North America. Its better to fight for those issues you agree upon than fight over what you disagree on! And everyone agreed that it was more fun to fight over more fish than less fish! Less fish is the result of lost habitat.

SESSION 2

A MANAGER'S VIEW OF FISH HABITAT

Session Chairperson: R.K. Lewis
Session Panellists: K. Edyvane
R.H. Winstanley
P.D. Jackson
Rapporteur: G.M. Newton

CHAIRPERSON'S INTRODUCTION

R.K. Lewis

South Australian Department of Fisheries
GPO Box 1625
Adelaide SA 5001

Fisheries managers are required to assimilate various data sources into management plans that address:

- ecosystem maintenance;
- sustainable exploitation;
- social and economic demands aimed at maximising the benefits to individual, specific groups and the current generation of resource users, to the possible detriment of the community in general and future generations;
- competing uses for particular habitats.

Fisheries management has always been espoused as being based on sustainable development.

This intent has been reflected in almost all fisheries legislation e.g. the South Australian Fisheries Act which is directed towards:

- ensuring through proper conservation, preservation and fisheries management measures, that the living resources of the waters to which this Act applies are not endangered or overexploited;
- achieving the optimum utilisation and equitable distribution of these resources; and
- the Department of Fisheries "Mission Statement", which aims "To conserve living marine and freshwater resources and develop them on behalf of current and future generations."

It can be arguably stated that the era of modern fisheries management reflecting this intent has been the last 20–25 years. However, regrettably, the history of fisheries management over this period has not been the achievement of the intent/level of expectations. Until recently most of the research/data sets have had a single species/stock perspective.

Without an holistic/integrated/ecosystem-based approach two outcomes have resulted:

- Many species and stocks have been systematically overexploited through serial depletion, because of the tendency to concentrate on a specific component of the ecosystem or stock and only redirect attention when other components have become threatened, and a preoccupation with the problems associated with the original components.

- There has been a failure to present advice on an integrated/ecosystem basis and therefore to "educate" the community/industry/politicians to think on an holistic basis as well as on the traditional fisheries viewpoint.

These have resulted in:

- a failure by these sectors to recognise the interrelated dependence of each component of the system;
- the need for a greater commitment to the maintenance of the system's integrity;

- a failure to recognise the collective extent of serial depletion;
- the need for a greater commitment to combat loss of habitats etc through other factors such as pollution, urbanisation;
- the commonly held image of fisheries and fishers as exploiters/plunderers, rapers.

These have occurred despite warnings from relevant scientific/management sectors.

In recent years this situation has improved. With the greater/increased awareness of environmental issues by a wide cross section of the community (ie the "greening" of the world) the required holistic/integrated/ecosystem approach has achieved greater prominence, support and credibility.

This is reflected through initiatives such as:

- Draft National Strategy for Ecologically Sustainable Development—3 recommendations;
- Biodiversity;
- Resource Assessment Commission, Coastal Zone Enquiry;
- A National Strategy for the Conservation of Australia's Biological Diversity;
- Ocean Rescue 2000.

The challenge for fisheries managers is to recognise the changes that are occurring, and to assimilate and apply them, both in the management of fisheries as well as in the wider educative role for management. Our keynote speaker Stan Moberly, suggested activism. Whilst this may appear self evident and sensible, it may not be simple to achieve. This is because of:

- the need for greater financial and personnel resources (just look at the costing associated with the sustainable development proposals);
- the need to develop analyses and methodologies to handle other than single species data bases;
- the need to reconcile the views of those trained in the more traditional fisheries management methods compared with those advocating the holistic integrated ecosystem-based approach.

As an example, in the South Australian Department of Fisheries this very issue has recently been vigorously debated.

Our three speakers, one each from the marine, estuarine and freshwater areas, will present overview and case study data to illustrate these points.

AN ECOSYSTEM–BASED APPROACH TO MARINE FISHERIES MANAGEMENT

K.S. Edyvane

SA Department of Primary Industries (Fisheries)
GPO Box 1625
Adelaide SA 5001

Abstract

Large-scale, multiple-use management is an ideal vehicle to implement and develop a holistic, integrated, ecosystem-based approach to fisheries management. It is now widely recognised that fisheries management must comprise a subset or component of a broader management of the whole ecosystem. Because of the "connected" nature of the marine environment, marine ecosystem management must address both system-oriented strategies, to prevent harm from pollution and overuse, and site-based strategies to protect habitats or to allocate and separate conflicting use. Large-scale, multiple-use managed areas, such as marine protected areas (MPAs), provide an ideal tool for implementing such an ecosystem approach to fisheries management.

The role of the 'fisheries manager' should be to provide input into: (i) the broad strategic approach to ecologically sustainable management, which will involve the use of environments and natural resources on a regional scale which matches the scale of marine ecosystems, and (ii) tactical habitat management, which will address a range of specific objectives such as biodiversity preservation, research, education and recreation, in addition to fisheries management.

The goal of the 'fisheries manager' should be to ensure the sustainable utilization of species and ecosystems. Inevitably, this will be linked with the maintenance of essential ecological processes and life-support systems, and also, the establishment of research and monitoring programs to monitor the effectiveness of management strategies. The challenge for 'fisheries managers' will be to redefine and broaden their role as 'habitat managers' within a new, integrated, ecosystem approach to management.

Introduction

In recent years there has been a new ecosystem-based focus in the field of natural resource management. This has arisen primarily from the continued decline of our natural resources despite massive regulatory efforts, and the recognition that there is a need to sustain ecosystems, in addition to the resources they produce (Kessler et al. 1992). In this new approach, ecological processes are given value and importance beyond the traditional commodity and amenity uses of ecosystems. These ecological processes include the provision and maintenance of a wide range of ecological "services", from climate regulation, protection from erosion, nutrient storage and cycling, pollutant breakdown and absorption, to, in terrestrial ecosystems, soil production and the maintenance of hydrological cycles (ie. groundwater recharge, watershed protection, and the buffering against extreme events). These ecological "services" and functions not only produce and sustain our natural resources but also underpin the quality of our life and our economy. In terrestrial ecosystems, this philosophy has been embodied within the concept and practice of "total catchment management".

In marine resource management, as in terrestrial resource management, there is also a growing awareness of the need to adopt a holistic, ecosystem-based approach to the management of our marine resources (Commonwealth of Australia 1991a). Despite regulatory efforts by individual agencies, pollution from point and diffuse sources, overfishing, loss of habitat from urban growth and coastal developments, and conflicts between competing user-groups (ie. fishing, aquaculture, tourism, recreation and conservation groups) continue to threaten our marine habitats and fisheries (see Commonwealth of Australia 1991b). Further, the greater degree of "connectedness" in marine ecosystems, compared to terrestrial ecosystems; the great mobility of organisms; and the extraordinary ability of water to transport both substances and organisms, result in the activities of one user group being more likely to directly and indirectly affect the activities of another (Kelleher and Kenchington 1991). For these reasons there is a greater need for integrated management of marine ecosystems.

In order to sustain our marine resources, it is now increasingly being recognised that fisheries management must be considered a component or subset of multiple-use, whole ecosystem management (Commonwealth of Australia 1991a). In this ecosystem-based approach to management, fisheries management is integrated and coordinated with the management of other uses and activities such as wastewater, shipping, tourism, recreation, conservation, mining, and industrial uses. While sectoral tasks and management to some extent may still use traditional methods, the real challenge in this new approach lies in ensuring effective intersectoral management. One of the greatest challenges for fisheries managers as we approach the next century will be to redefine and broaden their role as 'habitat managers' within this new framework of marine habitat management.

Developing a marine ecosystem management framework

Proponents for the sustainable use of marine resources have long recognised the need for the protection and maintenance of essential ecological processes. For instance, the World Conservation Strategy in 1980, clearly identified the preservation of life support systems as one of the four key elements in its global survival strategy. The four elements of this strategy include:

- the maintenance of essential ecological processes and life-support systems;
- the preservation of biological diversity at all levels, from ecosystem to genetic diversity;
- the sustainable utilization of species and ecosystems; and
- the establishment of research and monitoring programs to monitor effectiveness and environmental and global change (IUCN/WWF/UNEP 1980).

In translating this conservation strategy to our oceans and coastal ecosystems, we must recognise the strong linkages between the land, the ocean and the atmosphere, as well as recognising the "connectivity" of marine ecosystems. As such, land-based activities greatly influence the state of our coastal regions and its resources. At a global level, almost 80% of marine pollution is derived from land-based sources, with direct discharges accounting for some 44% and approximately 33% entering the marine environment as atmospheric inputs (GESAMP 1990). For this reason the conservation and management of marine ecosystems and their resources must ultimately entail effective management of land-based activities. At a global level this concept of integrated management of the land:sea boundary is known as Integrated Coastal Zone Management (ICZM).

As stated in the World Conservation Strategy, a management strategy for conserving and managing ecosystems must include not only protection of the diversity of life, but also the essential ecological processes and life-support systems which support it. Hence, for marine ecosystems, an integrated management strategy must comprise conservation of the attached fauna and flora of the seabed; water quality; the fauna and flora which live in the water column; and the key ecological processes (such as currents, tides, etc.) which sustain the ecosystem. Management of a seabed by itself without addressing pollution or overfishing will have little effect in the conservation or sustainable use of marine ecosystems. As such, an integrated management framework for natural ecosystems must comprise two essential components:

(1) general *ecosystem* protection to prevent harm from pollution and overuse; and

(2) *site-based* protection to protect habitats or to allocate and separate conflicting uses (Kenchington 1990).

For the management of marine ecosystems and their resources these components translate into two approaches:

(1) a broad strategic approach to ecologically sustainable use and management of environments and natural resources on a scale which matches the scale of marine ecosystems; and

(2) tactical site or marine habitat management to address specific objectives of biodiversity preservation, research, education and recreation.

Measures which seek to address the tactical objectives of habitat management without the broader strategic framework are likely to fail to address the broad requirements of conservation and the sustainable use of marine ecosystems (Kenchington 1990). For an integrated approach to fisheries management there is a need to include the essential components of ecologically sustainable use and also, recognition of the need for ecosystem management. The recent report released by the Ecologically Sustainable Working Group for Fisheries (Commonwealth of Australia 1991a) clearly identified these objectives and further, recommended a number of key steps to achieve the ecologically sustainable use of our fisheries resources. Many of these recommendations are useful in formulating a set of guidelines for achieving an ecosystem approach to fisheries management (see Table 1).

Multiple-use, marine protected areas and fisheries management

In recent years there has been a dramatic shift in the role of marine protected areas (or MPAs) in fisheries management. Until recently MPAs were seen primarily as sites for the protection of 'critical habitats' of economically important species. Estuaries and wetland habitats, such as seagrasses and mangroves, were included because they protected key parts of the life history of species. Specific fisheries were enhanced by the protection of identified nursery areas, feeding areas and spawning areas. MPAs however, can provide a number of other important roles in fisheries management, in addition to critical habitat protection. These include: areas for stock replenishment, ie. 'harvest refugia'; areas for monitoring the natural fluctuations in stock; and areas for resolving conflict between competing users of marine resources and habitats. In the latter regard, MPAs are increasingly being seen as a vehicle for implementing an ecosystem-based approach to fisheries management.

The essential tool of multiple-use, ecosystem-based management in MPAs is the zoning of human uses and activities on a geographical basis. Uses such as fishing, tourism, recreation, conservation and maritime shipping, are essentially coordinated and integrated through the development of a zoning plan. Zoning not only provides a mechanism for reducing conflict between competing user groups but it also provides a mechanism for the effective protection of 'critical areas' through the creation of 'buffer zones'. Probably the most well-known example of multiple-use management of a marine ecosystem in Australia is the Great Barrier Reef Marine Park in Queensland. Within this 348 700 sq.km park, fishing and a wide variety of other human uses (including tourism, recreation, preservation and scientific research) are managed on an ecologically sustainable basis through six types of zones within the marine park (GBRMPA 1985). All activities are both managed and coordinated on an integrated basis by a single regulatory authority, the Great Barrier Reef Marine Park Authority.

Zoning of uses is a viable approach to marine fisheries management that deserves serious evaluation. Not only is it an ideal tool for implementing an ecosystem approach to marine resource management, but it also has the potential for increasing consistent sustained harvests through the creation of 'harvest refugia'. Further, zoning also allows reduced fisheries regulations and thus can simplify enforcement and compliance. It may also allow dynamic market forces to optimize harvest sizes and seasons, and may permit those same forces to drive development of more efficient nondestructive fishing gear (Davis 1989).

While multiple-use zoning has yet to be implemented in many regions and countries, habitat management is increasingly being recognised as an essential component of fisheries management, in addition to the traditional single-species approach to management. This is evidenced by a recent proposal in 1990 to establish large fishery reserves off the Atlantic coast of the South-Eastern United States. In this proposal, reef fishes would be managed by both conventional means of size and bag limits, and by the establishment of a set of marine reserves, where reef fishing would be prohibited (Huntsman and Vaughan in press). The proposed marine reserves would encompass some 20% of the region's reefs.

Large-scale zoning is also of immense value in 'adaptive management'. Management plans can use zoning as a research tool to establish a framework for scientific testing of concepts, methods and assumptions. As such, large-scale ecological experiments can be conducted through the zoning mechanism to investigate the effects of a particular management regime. For instance, in certain parts of the Great Barrier Reef Marine Park, zoning regulations are presently being used to establish the ecological effects of particular fishing methods. In this self-regulatory approach to fisheries management, the management regime is adjusted and regulated through the results of monitoring.

In recent years the establishment of large, multiple-use managed marine areas in Australia has received greater attention with the announcement of a national, 10-year marine conservation program called 'Ocean Rescue 2000'. One of the essential elements of this program is the establishment of a national, representative system of MPAs. While some states in Australia (such as Queensland and Western Australia), have established several large, multiple-use marine parks, some states such as South Australia have yet to establish large, multiple-use managed areas (Table 2). In these States, continued funding under 'Ocean Rescue 2000' will be critical in establishing such areas.

In summary, the increasing importance of MPAs as tool for fisheries management is a sign of a new ecological order for fisheries management. This recent change in focus stems primarily from the failure of traditional single-species based management practices to halt the decline of fisheries resources. More than ever, MPAs are increasingly being seen as a vehicle for a new order of ecosystem-based management of fisheries, rather than the traditional single-species approach to management. The challenge for 'fisheries managers' will be to redefine and broaden their role as 'habitat managers' within this new, integrated, ecosystem approach to management.

References

Commonwealth of Australia (1991a). Ecologically Sustainable Development Working Groups Final Report—Fisheries.

Commonwealth of Australia (1991b). *The Injured Coastline*. Report of the House of Representatives Standing Committee on Environment, Recreation and the Arts (HORSCERA).

Davis, G.E. (1989). Designated harvest refugia: the next stage of marine fishery management in California. CalCOFI report, Volume 30.

GESAMP (IMO/FAO/UNESCO/WMO/WHO/IAEA/UN/UNEP Joint Group of Experts on the Scientific Aspects of Marine Pollution) (1990). The state of the marine environment. *Rep.Stud.GESAMP* 39:111pp.

Gilmour, A.J. and J. Connor (1991). A proposal for legal and institutional alternatives for the consideration and management of Australia's maritime ecosystems. Paper prepared for the Australian Conservation Foundation and the World Wide Fund for Native Australia. Melbourne, June.

Great Barrier Reef Marine Park Authority (1985). *Zoning the Central Section*. Townsville, Australia.

Huntsman, G.R. and D.S. Vaughan (in press). Relative risks in managing reef fishes between marine reserves and size/bag limits. *Proc.World Fisheries Conf.* 1992.

IUCN/UNEP/WWF (1980). A Strategy for World Conservation. IUCN/UNEP/WWF, Gland, Switzerland.

Kellaher, G. (1991). Cost recovery for fisheries, Submission to the Industry Commission draft report. *Cost Recovery for Managing Fisheries*, Great Barrier Reef Marine Park Authority, Canberra.

Kelleher, G. and R. Kenchington (1991). *Guidelines for Establishing Marine Protected Areas*. IUCN, Gland, Switzerland.

Kenchington R.A. (1990). *Managing Marine Environments*. Taylor and Francis, New York.

Kessler, W.B., H. Salwasser, C.W. Cartwright and J.A. Caplan (1992). New perspectives for sustainable natural resource management. *Ecol.Appl.* **2**, 221–225.

McNeill, S. (1991). The Design of Marine Parks with an Emphasis on Seagrass Communities. MSc thesis, Macquarie University.

Table 1. Guidelines for achieving an ecosystem approach to fisheries management (adapted from ESD Fisheries 1991)

An Ecosystem-Based Approach to Fisheries Management

Goals

- Recognition of fisheries management as a subset or component of ecosystem management.

- Adoption of ecologically sustainable use and inter- and intragenerational equity as goals of fisheries management:

    - sustainable utilization of species and ecosystems;

    - maintenance of essential ecological processes and life-support systems;

    - the preservation of biological diversity at all levels, from ecosystem to genetic diversity.

Management

- Establishment of Marine Protected Areas (MPAs) involving large-scale, regional management of multiple-use areas for:

    - resolving conflict by competing user groups through zoning activities;

    - protection of biodiversity;

    - enhancement of fisheries management, through refuge sites for stock replenishment, and protection of fish nursery areas.

- Adoption of adaptive and flexible management methodologies (such as Adaptive Environment Assessment and Monitoring), which recognise the uncertainty associated with resource management of biological systems.

- Promotion of habitat amelioration and enhancement (adopt principle of 'no net loss' of habitat).

- Conservation of both 'critical' and 'ecologically representative' habitats for fisheries management.
- Management of environments and natural resources on a regional scale which matches the scale of marine ecosystems, for instance:
  - Marine Protected Areas (MPAs);
  - National Maritime Authority (proposed by Gilmour and Connor 1991);
  - Coastal Zone Management Authority (proposed by Kelleher 1991).
- Development of an integrated management framework:
  - 3–5 year strategic management plans for *all* fisheries (including critical habitats and ecological processes, potential threats, and performance and sustainability criteria);
  - greater community and industry involvement in decision-making processes;
  - management of fisheries on a multi-species level;
  - initiation of intersectoral management mechanisms:
    * regional and State "ecosystem" committees (to address intra-sectoral and intersectoral issues), in addition to traditional single-species fishery committees;
    * use large-scale, multiple-use MPAs as a tool for fisheries management and vehicle for ecosystem management.

Research
- Identification and conservation of the critical ecological processes and habitats which sustain fisheries, eg. upwellings, environmental "cues", nursery areas, feeding areas, breeding areas, "sinks" or "sources" for larvae.
- Identification of the spatial and temporal scale of the critical ecological processes.
- Assessment of the impacts of pollution, fishing, aquaculture and fishing methods on critical ecological processes and habitats (ie. assess ecosystem and habitat integrity).
- *Separation* of the effects of overfishing from the effects of habitat degradation (a combination of both?) by identifying:
  - relationship of fish to habitat;
  - effect of environmental influences on habitat.
- Determination of the need, purpose, location, design and size of marine protected areas and their role in maintaining a particular species and/or aquatic ecosystem.
- Development of more robust predictive tools for stock assessment, environment assessment and bioeconomic analysis.
- Development of data information/retrieval systems such as GIS (Geographical Information System).
- Development of predictive, adaptive, GIS-based, multi-species models to manage fisheries (eg. AEAM (Adaptive Environmental Assessment and Management) and ecosystem-based models).
- Development of biological and ecological criteria to assess the goals of sustainability and "ecosystem health", with statistical power to detect effects of unsustainable use.
- Development of economic criteria to assess sustainability.

Table 2. Marine protected areas (MPAs) in Australia, up until 31 May 91 (adapted from McNeill 1991). Figures include the number of MPAs in each State/Territory; total area of MPAs in each State/Territory; area of MPAs as a percentage of total State/Territory waters; area of MPAs as a percentage of total area of MPAs in State and Commonwealth waters; area of MPAs as a percentage of total area of MPAs in State and Commonwealth waters minus the area of the Great Barrier Reef Marine Park (GBRMP)

State/Territory	No.	Area km2	% Area State	% Area Total	% Area Total -GBRMP
Queensland	139	354 799	24.5	90.5	21.2
External Territories	4	17 975	–	4.6	38.0
Western Australia	19	14 328	20.3	3.6	30.4
Northern Territory	10	2 841	8.2	0.7	6.0
New South Wales	22	924	8.7	0.2	2.0
Victoria	27	537	5.4	0.1	1.1
Tasmania	17	488	2.8	0.12	0.6
South Australia	28	295	1.4	0.07	0.6
ACT	4	8	4.0	0.002	0.02

ESTUARINE ISSUES FROM THE MANAGER'S VIEWPOINT

R.H. Winstanley

Marine Science Laboratories, Department of Conservation and Environment
PO Box 114
Queenscliff VIC 3225

Introduction

The ecological value of estuaries and their importance to fisheries have been appreciated for many years. So too has the array of fishing and environmental factors—in the estuaries, their catchments and adjacent coastal waters—that impacts on estuarine fish stocks and ecosystems. Long before the term "ecologically sustainable development" was coined, the concept of "ecological sustainability" of estuaries and their fish resources was well established amongst management agencies, fishers and the wider community.

From a background where the focus was on the exploited fish stocks, management attention has shifted towards managing fisheries in the broader context of estuarine living resource conservation. Clearly, this can only be achieved through understanding the ecological linkages between fish stocks and estuarine habitats and the impacts of polluting and disturbing factors on those habitats and stocks.

In the past, all of this often seemed too hard, the information base too small and the ecological enormity of the task too daunting. The political balance appeared too heavily weighted in favour of commercial activities and economic development goals, most of which are fundamentally incompatible with the long term maintenance of estuary habitats and fish stocks.

In the fields of fisheries management (and here I include aquatic conservation), the manager is the link between the scientists and the politicians in the decision-making and planning processes.

This can be a difficult place to be in any field. It is a difficult place to be in fisheries management. It is a particularly difficult place to be in relation to estuaries where the range of competing interests and pressures operating on the fish resources and their habitats are so numerous and complex.

In this environment, manager must ensure that research effort is focused on the critical long term and short term information-gathering programs.

Without understanding the continuing difficulties, there are a number of factors which have changed in a way that promises real hope that the ecological basis for estuarine fish stock conservation will be achieved.

Where are we today?

Karen Edyvane has outlined the various factors which impact on estuarine (and other) habitats and, over the next two days, we will hear of the importance of estuarine fish stocks and habitats, so I will not detail them here.

In the 1990s, we are mainly dealing with highly modified estuaries. This situation arises from:

- our history, which saw estuaries and marine embayments as the sites of our earliest settlements, and the intensification of population concentrations and associated aquatic impacts ever since; and
- the fact that estuaries are where the impacts of mismanagement in the catchments and hinterland converge.

In addition, while those of us in fisheries agencies have been attending to long-standing problems, we have been overtaken by new factors such as the deliberate development and usage of products (e.g. tributyl tin antifouling paints) and industries (e.g. ecotourism) and the inadvertent spreading of exotic organisms. In some cases, introduced organisms have come to dominate estuarine communities; in other instances, exotic toxic algae have threatened aquaculture and shellfish harvesting industries.

Consequently, fisheries managers have struggled to achieve the proper recognition and effective conservation of estuarine habitats.

We lack basic information on:

- the ecology of fish species at every stage of their life history;
- the critical habitat requirements of these stages;
- the distributing or polluting factors which impact or threaten to impact on those habitats.

This has weakened our ability to put credible and successful cases to ensure that development activities proceed in ways that do not compromise estuarine fish stocks and habitats. The onus of proof in Environmental Impact Statements (EIS) and related planning processes continues to fall on aquatic resource managers—not on developers.

In many instances, the EIS and planning processes are taking place without early input from aquatic resource managers. As our keynote speaker has already pointed out, we must become more assertive and proactive and communicate our message more effectively to ensure that estuarine fish and habitat conservation issues are placed firmly on the agenda of all of the agencies involved in these processes.

We are often constrained by problems of overlapping and fragmented jurisdictions and responsibilities, and the absence of clear overall aquatic conservation strategic plans; this applies at both state/territory and federal levels. Sometimes, we are further constrained by deficient legislation which does not provide the powers needed to underpin effective habitat protection.

For example, in Victoria, we have the Fisheries Act providing the powers for commercial and recreational fisheries management; we have the Flora and Fauna Guarantee Act providing the powers to protect designated species or communities from threatening processes; we have the Environment Protection Act providing the powers to maintain water quality standards; and we have the National Parks and Land Acts providing the powers to manage public land. However, there is a large hole in this legislative combination which severely limits our broad marine and estuarine conservation powers.

In short, fisheries and aquatic conservation managers frequently lack both the information and the authority to properly protect the fish and aquatic habitats for which they are responsible.

Too often, managers and interest groups have been preoccupied with symptoms rather than causes of fish habitat problems. In some instances, anglers push artificial reef proposals while agencies struggle to protect threatened productive natural reef habitats.

Too often, governments are facing up to problems only when they have deteriorated to crisis point, for example:

- the recent focus on national blue-green algal blooms in inland and estuarine waters;
- improvements to Sydney's sewage treatment and disposal;
- responses to public health scares over Georges River oysters.

In each case, the problems have been known for years. In instances like the current plan for treating the eutrophication problems of the Peel-Harvey Inlet, the costs of belated treatment of the symptoms are huge.

Fisheries managers still face the widespread and dangerous misconception that, provided we establish representative Marine Protected Areas and protect designated threatened species and communities, we are doing a thorough job of habitat conservation.

Government responses

In the face of these challenges and impediments, state and territory fisheries agencies have increasingly focused resources on estuarine habitat research and management programs. For example, the NSW Fisheries biennial report of fisheries indicates a large commitment to this area during the 1989–1991 period.

Increasingly, fisheries agencies are giving prominence to aquatic habitat protection as a cornerstone of their corporate plans (e.g. SA Department of Fisheries) and strategic research plans (e.g. NSW Fisheries).

To assist with planning and management at all levels of government, information and guidelines for the conservation of estuarine habitats are being produced.

Managing agencies are producing estuary-specific management plans which:

- recognise the full range of community benefits arising from the estuaries;
- consider the cumulative impacts of polluting and disturbing factors;
- involve all interested agencies and groups;
- are supported by legislation.

As an aside, it is interesting to note that what has been happening nationally is mirrored in this Society: the shift from a focus on fin-fish and their biology, to a wider range of aquatic species, to fisheries issues, and now to fish and aquatic habitat conservation—hence this workshop! Clearly, this professional society has a significant role to play in leading agencies, governments and community responses to these issues.

Wider community responses

On a wider front, there are a number of promising signs.

We have noted the upsurge in community concern about marine and estuarine conservation issues in the last few years.

Commercial fishing industry organisations are moving away from reacting to specific environmental issues (e.g. coastal discharges, pulp and paper mill proposals) towards proactive campaigning for fish habitat protection. In NSW, the industry has established Ocean Watch to advise it and to act as its advocate on marine conservation issues. Similarly, the Queensland Commercial Fishermen's Organisation has engaged professional assistance to promote fish habitat protection, particularly in estuaries.

Recreational Fisheries bodies are increasingly taking an active part in water and land management reviews and in marine and estuarine planning processes.

Broad strategic planning frameworks

We should take further encouragement from the increased attention to broad marine and coastal strategic planning.

For example, at a national level, the Ocean Rescue 2000 program and Resource Assessment Commission coastal zone enquiry will provide a broad framework for the future of marine conservation and coastal management planning.

At the state level, the Victorian Land Conservation Council, a public land management planning body, is preparing a strategic planning framework for coastal management and marine conservation in Victoria.

The promise of the national Ecologically Sustainable Development program is that it will encourage and lock in all levels of government and all areas of public and private enterprise into a set of principles that place due emphasis on the conservation of estuarine resources.

If we look at current national initiatives in areas such as Landcare, salinity control and catchment management, we see references to "holistic approaches", "whole ecosystem management" and "total catchment management". This indicates that there is a widespread convergence occurring in the mindset and goals of all those involved in the usage and management of natural resources across Australia. By adopting proactive strategies, fisheries managers should be able to capitalise on the momentum of this movement to make new allies in addressing the task of protecting fish habitat.

In conclusion, many people are sceptical and these current aquatic and wider conservation initiatives may prove to be mainly rhetoric—or passing fads. We must recognise that no habitat protection policies and strategies can be relied upon to have the desired effect if they are based on inadequate knowledge or if they lack total commitment by all parties.

FRESHWATER HABITAT PROTECTION— A MANAGER'S PERSPECTIVE

P.D. Jackson
Fisheries Division, Queensland Department of Primary Industries
GPO Box 46
Brisbane QLD 4001

Introduction

Many of Australia's inland aquatic habitats are seriously degraded, some to the point where their long-term ecological sustainability must be in serious doubt (Hart 1992). From a fisheries viewpoint, habitat degradation has been implicated in the decline of many native species. The Freshwater Fish Action Plan (Wager and Jackson in press) lists habitat degradation along with negative interactions with introduced species as the major threatening processes for freshwater fishes (see also Jackson, Koehn and Wager this meeting). Fish under threat range from small species of high conservation value such as the honey blue-eye (*Pseudomugil mellis*) to important recreational species such as Murray cod (*Maccullochella peelii*).

This paper looks at the protection of fish habitat from a manager's perspective and uses the plight of the Mary River cod (*Maccullochella sp.*) to illustrate some of the points.

Managing inland waters

Managing habitat for long-term sustainability of fisheries in inland waters is inevitably about managing competing uses of resources within a catchment (e.g. instream *vs* offstream use of water, land use for agriculture *vs* retention of catchment vegetation etc.). Protection of habitat cannot be achieved without consideration of land use practices within the catchment. In only very few circumstances, perhaps headwater tributaries in forested areas or catchments contained wholly within National Parks, will there be no change in aquatic habitats. Multiple use of catchment resources will inevitably mean change.

Given the above, the data required by managers may be summarised as follows:

- *What are the most important areas of habitat to protect?*
  Baseline data must be available on existing habitats and their fish communities. Are there priority habitats to be protected? Are there particular tributaries within a catchment that are more important than others? This may be important if there are multiple choices for a proposed impoundment site for example.

- *What are the key characteristics of the habitat?*
  It is important to determine the habitat variables that are predictive of fish community structure. What are the habitat characteristics that a manager must try to protect?

- *What are the trends?*
  Is the habitat stable or is it degrading?

- *What are the acceptable levels of change in habitat variables?*
  What are the boundaries of acceptable change? For example, how much water can be harvested for offstream use and how much can the seasonality of flow regimes be altered without major impacts on fish communities.

- *What are the quantitative relationships between habitat change and catchment activity? What are the causal factors in habitat degradation and how can they be managed?*

  It is important to establish the linkages between catchment activities and instream habitat. Inevitably quantitative relationships will have to be established. For example, what land use practices contribute to sediment runoff and how can land use practices (at the on-farm scale) be modified to reduce sediment contribution to an acceptable limit?

- *What are the habitat rehabilitation options?*

  In certain circumstances management actions to rehabilitate degraded habitat may be both desirable and practical. For example, replanting riparian vegetation may be an option or it may be feasible to rehabilitate the channel form to create appropriate habitat diversity (e.g. pools and riffles etc.) in streams with reduced water flows. A manager needs data on specific rehabilitation requirements.

The Mary River cod — a case study

The Mary River cod (*Maccullochella sp.*) is an important recreational angling species that has greatly declined in both numbers and distribution since the early 1900s. It is thought to be identical to a fish that previously occurred in the Brisbane, Logan, Albert and Coomera Rivers in south-eastern Queensland. It is currently restricted to a few larger tributaries of the Mary River including Obi Obi, Six Mile, Tinana and Coondoo Creeks. It is generally found in deeper pools of these relatively undisturbed tributaries where fallen timber, branches and boulders provide cover.

The Mary River flows through a multiple land use catchment after rising in the Conondale Ranges north of Brisbane. The catchment area is about 9595 km^2 of which approximately 65% is freehold agricultural or grazing land. The remainder is predominantly State forest with a significant amount of this being exotic pine. National Parks account for less than 1% of the total catchment area. There are water storage dams on three of the tributaries and further water harvesting is planned.

Although definite data are not available, threats to Mary River cod appear to include: dams and weirs as barriers to movement, loss of native riparian vegetation, extensive siltation from catchment erosion due to land use practices, stream channel damage due to land use practices, steam channel damage from sand and gravel extraction, possible competition with translocated species (golden perch, *Macquaria ambigua* and silver perch, *Bidyanus bidyanus*) and overfishing.

Available data

In 1991, habitat surveys of the Mary River were undertaken by staff from the Queensland Fisheries Divison's Southern Fisheries Centre. These surveys have provided 'broad brush' information on the condition of instream habitat and disturbance types within the Mary River catchment. Additional habitat information and fish community structures in the Mary River have been obtained by Brad Pusey from the Centre for Catchment and Instream Research, Griffith University.

Data are available on land use within the catchment and some information is available on the status of riparian vegetation from aerial photographs obtained by the Queensland Forest Service.

Information on Mary River cod distribution and abundance remains largely anecdotal or is derived from catch records by local anglers.

Data requirements

Clearly there are gaps in available knowledge that must be filled if Mary River cod habitat is to be effectively managed to ensure the long-term conservation of the species.

Definitive data must be obtained on the present distribution of cod in the Mary River system together with more concise data on the habitat requirements of the fish. The Queensland Fisheries Division has received funding from the Australian National Parks and Wildlife Service to undertake the necessary fish surveys this year. The information obtained will also provide baseline data on cod distribution and abundance and, together with future surveys, will enable trends to be evaluated.

The links between instream habitat change and catchment activities must also be established. Ultimately management prescriptions will need to act at the individual farm level if they are to be effective in the Mary River catchment.

Finally there is a need to establish the acceptable levels of habitat change. Most pressing is the need to determine environmental flow requirements in the tributaries containing cod, e.g. Six Mile Creek and Obi Obi Creek already have impoundments on them and more water harvesting is planned.

Management options

Effective protection and management of Mary River cod habitat cannot be achieved without a whole catchment approach. The Mary River catchment is to become a pilot catchment for the Queensland Department of Primary Industries' Integrated Catchment Management Initiative and a community driven catchment coordinating committee has already been established. Ultimately a catchment management plan will be produced and the protection of instream habitat will be part of that plan.

Data obtained on specific habitat requirements of Mary River cod and effects of catchment activities on cod habitat will enable effective management measures to be put in place. However, it is important to recognise that managers cannot always wait until definitive data are available and often must act on the best information to hand at the time.

In the case of the Mary River, options for future impoundment sites are being considered now and preliminary data on cod distribution must be used to ensure proper input to the initial planning process. Similarly, some management options can be implemented immediately. A good example is the protection of remaining native riparian vegetation and the implementation of measures to revegetate stream banks.

Conclusions

Ultimately, managers require specific information on habitats, what are the key habitat variables and what are the limits of acceptable change. In many cases these data are not yet available but often general information is available on the broad threats to habitat (e.g. removal of riparian vegetation, increased sediment input, regulation of stream flows). Managers must use the best available data to mitigate these effects whilst encouraging and supporting habitat requirement research.

References

Hart, B.T. (1992). Ecological condition of Australia's river. *Search* **23**, 33-37.

Jackson, P.D., J.D. Koehn and R. Wager (this meeting). Australia's threatened fishes 1992 listing—Australian Society for Fish Biology.

Wager, R. and P. Jackson (in press). The Action Plan for Australian Freshwater Fishes. *Australian National Parks and Wildlife Service*, 128pp.

DISCUSSION OF SESSION 2

Recorded by G.M. Newton

Bureau of Resource Sciences
GPO Box 858
Canberra ACT 2601

The three panel presentations were each followed by a short discussion, following which the Chairperson, Rob Lewis, opened the meeting for more general discussion.

The first speaker, Karen Edyvane of the South Australian Department of Fisheries, spoke of fisheries management as a subset of ecosystem management. She emphasised the need to look at critical ecological processes and their temporal and spatial scales, as well as to assess impacts on these processes by the various user groups of the marine environment. Monitoring criteria and indicators were seen as fundamental considerations. Greater use of information technology, Geographical Information Systems (GIS), and predictive tools was encouraged.

John Glaister commented on *Karen Edyvane's* overhead transparency on management of the marine environment, which advocated another level of bureaucracy e.g. the Australian Fisheries Management Authority (AFMA), as the single responsible agency. Wasn't that flying against the principle of collective wisdom?

Karen Edyvane justified her viewpoint by suggesting a model for debating the issues may be through the formation of a single ecosystem committee. This committee could be formed by community representatives of the various user groups who would then discuss the issues related to fisheries and potential conflicts.

John Glaister maintained that habitat is an integral part of fisheries management and rather than move into a new area we need to focus on current real issues.

Karen Edyvane however, pointed to the example provided by the Great Barrier Reef Marine Park Authority, which deals with a multiple use environment, requiring other user groups to take part in the decision-making process. We need to broaden our view!

The second panelist, *Ross Winstanley*, spoke about the need for a better understanding of fish stocks dependent on estuaries. Barbara Richardson commented that currently the onus of *proof* of habitat issues rests with fisheries scientists and managers. The topic is of such importance that there needs to be a change of onus, with involvement by other disciplines in decision-making about saving habitat, with consequent better management.

Ross Winstanley agreed that a wider forum is needed. For example, decision-making in Port Phillip Bay covers deepening of shipping channels and dumping of sediment. Land disposal is controlled by the Environmental Protection Agency, but attention needs to be given to effects underwater e.g. in dips and hollows.

Rob Lewis commented that part of the answer lies in mobilising all sectors of the community so that they collectively own the initiative.

Following *Peter Jackson's* panel presentation, which focussed on the data requirements of managers of freshwater habitats, Campbell Davies suggested that because of the lack of data relevant to survival of Mary River cod, the onus should rest with the developers rather than the scientists, to show that their activities will not threaten the species. Peter Jackson responded that there is now little water remaining in the Mary River, and the onus is on water resource authorities to facilitate the relevant research by fisheries scientists.

The Chairperson, Rob Lewis, led the *general discussion*, which concentrated on issues of research and assessment, on education and communication, and on mobilisation of the community, by referring back to John Glaister's earlier question about another level of authority. Formal management committees e.g. boards of directors for South Australian fisheries, have specific responsibility and accountability, but there is a greater need for holistic, integrated, ecosystem approaches, a need for new rather than traditional mechanisms. We need to identify the advice required by fisheries managers—onus of proof, ownership, avoiding habitat change etc—but we do not have the bio-economic models needed to place economic values on fish stocks and habitats. The importance of integrative catchment management overrides the State perspective.

Peter Young commented that over the last five or so years greater significance has attached to other stakeholders—agriculturalists (including aquaculturalists), mining, forestry, conservation, tourism etc as well as fisheries interests. Historically, decision-making has been by fisheries managers rather than environmental/conservation departments, but with the current focus on ecosystems, fisheries managers need to reassess their role—and quickly! It is essential in the aquatic environment that fisheries managers remain in control of the agenda.

Karen Edyvane cautioned the need to put the issues in perspective when considering other stakeholders. She also highlighted the fact that about 80% of decisions are influenced by local governments and therefore may be out of our hands.

Stan Moberly questioned where fisheries sit in the hierarchy. They should be involved in the strategic planning with influence equal to irrigation, mining, forestry etc, not just responding to Environmental Impact Statements on, for example, endangered species issues.

John Koehn saw the management of threats to habitat—direct or indirect—as requiring links with other agencies. The Flora and Fauna Guarantee Act of Victoria (1988) provides for listing of threatened species, and communities, and for threatening processes.

Ross Winstanley stressed that the process is dependent on the availability and quality of scientific information, on being able to distinguish clear viable links between fish, habitats and processes. In marine coastal waters, threats are still being responded to on a reactive basis without the equipment to utilise legislation. In Victoria, the Fisheries Act precludes direct exploitation, there is a mechanism for marine parks, the EPA controls water quality, but the legislation does not provide for broader aquatic conservation.

Peter Jackson was concerned as to whether, although the Act recognises threats, it is effective in practice. Does the action plan become implemented or change anything? In the absence of existing data, the onus must be on the people affecting the habitat to prove it will not be altered or destroyed.

Karen Edyvane pointed out that a recommendation of the Ecologically Sustainable Development (ESD) working group was that potential impacts be identified so that breaches of the management plan can be legislated against. John Koehn urged that priority should be given to identifiable threats. For example, dams pose many threats, but are allocated few resources, and consequently little management.

Hugh Cross referred to the stakeholder, often at the top of the organisation, who allocates the funds, and who needs to be convinced of the value to the community, as well as to ecological and catchment processes.

Rob Lewis wanted to know how we can improve valuation of habitat in economic terms, which is always a dilemma for fisheries managers. Without a quantifiable value/figure, there is a danger of lost credibility in the face of figures supplied by developers and engineers, as their figures may be shaky but are seen as solid.

Jenny Burchmore's experience was that the bio-economic models can be a trap. For example the value of a fishery is expressed as an *annual* loss in an ESD document, but a sustainable fishery has an infinite value.

Duncan Leadbitter reminded the meeting that the tools are available for implementing habitat management in the form of the EIS (Environmental Impact Statement), but the EIS doesn't contain much science. What is needed from scientists are appropriate guidelines, especially on pre-impact studies, focussing more on environmental needs than developers' wants.

Rob Lewis agreed, but the time-line on EIS's is a constraint. Karen Edyvane was concerned that because drawing up of an EIS is the responsibility of one agency, fisheries managers get little input. An integrated approach is needed towards looking after the environment, and we shouldn't be too negative about strategic planning, with recommendations and guidelines at a higher level. Predictive modelling and simulation modelling can be important tools.

Terry Walker had difficulty with the concept of marine parks, where the complex issues cause public confusion, often because they are set up with inadequate information. When managers try to accommodate everybody, the outcome becomes irrational.

Karen Edyvane stressed that it wasn't an easy process. The value of large regional multiple use areas is the buffer they give to highly protected areas, for example a fish sanctuary.

Ross Winstanley supported Duncan Leadbitter on the need for a more assertive approach by fisheries managers in the marine and coastal planning processes.

Kruno Kukolic's experience from the ACT planning authority was of two sets of guidelines which have been created for the conservation of wetlands. These are supplied to consultants who can incorporate the principles in EIS.

Jenny Burchmore claimed that 80% of decisions are made by planners, with fisheries interests having little input. But fisheries managers tend to overlook fish habitat protection in their educational programmes. Other Government departments are often more of a problem than developers, because they lack an understanding of the issues. One remedy would be workshops for public works engineers and planners, and education packages directed at bureaucracy. However, the downside of all this could be that other departments may then want to take the initiative from fisheries managers. Karen Edyvane believed that the best way to educate planners is to work alongside them in the planning process, and to get the fish habitat workers of fisheries departments fully involved.

John Glaister wasn't opposed to the various stakeholders being involved in the management process. However, the fact that fisheries researchers are the source of information on fish habitat seems to have become devalued currency. Rather than creating larger bureaucracies to deal with issues germane to fisheries research, we need to re-establish our importance as the data source.

Rob Lewis's summary of the discussion of Session 2 was that the information supplied to fisheries managers should be improved, there should be a more integrated perspective, there is a need for resources, a need to identify all relevant groups, including other government instrumentalities, and a need to coordinate like interests.

SESSION 3

ORGANISMS AND ENVIRONMENTAL RELATIONSHIPS
I — CASE STUDIES

Session Chairperson: P.C. Gehrke
Session Panellists: I.R. Poiner
S. Swales
M.T. Bales
R.M. Morton
Rapporteur: G.A. Thorncraft

CHAIRPERSON'S INTRODUCTION

P.C. Gehrke

*NSW Fisheries, Inland Fisheries Research Station
PO Box 182
Narrandera NSW 2700*

Sustainability has become a buzz word for the 1990's, especially in the context of ecologically sustainable development, but what does it mean in the context of sustaining both the fishing industry and fish habitat? My dictionary informs me that to sustain an object or activity is to keep it from failing, to endure without giving way, or to support by giving aid. So sustenance means actually doing something to achieve longevity. Sustaining a fishery, then, ensures that the fishery is able to continue into the future, and fisheries management, regulation and research all contribute to the attainment of that goal. We are here to discuss how we might sustain fisheries through sustaining fish habitat, which begs the question of what we mean by sustaining fish habitat and how we might actually go about it.

It is naive to conduct a fishery in a pristine habitat and hope the habitat will remain unchanged in any way. When additional pressures from other human activities are imposed upon fish habitats, subtle and not-so-subtle habitat changes become apparent. To borrow an analogy from military field hospital terminology, habitats can categorised on a triage basis with respect to the severity of their injuries: (i) Pristine, undamaged habitats are nature's crown jewels which only need sustenance in the form of protection to ensure they are not damaged; (ii) Walking-wounded habitats are those that have been damaged in some way, yet still support a component of their original fish communities. These are the habitats which stand to gain most in fisheries production from active sustenance on a cost–benefit basis; (iii) Irrecoverably-wounded habitats, on the other hand, are beyond all hope of economical rehabilitation to support fish communities, although social pressures to restore the habitat may prevail. These habitats do not need sustenance, they need to be rebuilt.

So then, to sustain the habitats upon which Australia's fisheries are based, we do not seek to restore modified habitats to their pristine condition. Nor are we concerned with the irretrievably and perhaps repeatedly damaged habitats which presumably no longer support viable fisheries. We are concerned with protecting nature's family jewels, many of which have some natural protection of their own due to their isolation. Mostly though, we are concerned with sustaining the walking-wounded habitats, to prevent further damage and where possible, to improve their condition.

The interactions between organisms and their environment are complex, and vary both among species and among habitat types. The case studies we will examine in this session come from shallow coastal habitats, estuaries, rivers and lakes, and focus on changes caused by natural disasters, impoundment of waterways, eutrophication and real estate development. While each of these changes undoubtedly impacts upon fisheries, each also has a positive component which ameliorates the destructive aspects and even provides clear opportunities to recover damaged habitat. The important message shared by these cases is that all activities which change habitats exact a cost in fisheries production, but with attention to details, such as buffering processes and reconstruction, the cost can be reduced and even turned into a profit. The challenge for us as researchers and managers is to devise and implement strategies which not only sustain natural habitats, but which in turn sustain our fisheries.

MAINTAIN OR MODIFY— ALTERNATIVE VIEWS OF MANAGING CRITICAL FISHERIES HABITAT

I.R. Poiner, N.R. Loneragan and C.A. Conacher

CSIRO Division of Fisheries, Marine Laboratories
PO Box 120
Cleveland QLD 4163

Summary

Recently fisheries management in Australia has shifted to emphasise management of resources within the principles of ecologically sustainable development. This has resulted in management to sustain fish stocks, maximise economic efficiency when harvesting those stocks, and a trend towards granting property rights to the fishers. To achieve the goal of management to sustain fish stocks, a major focus of fisheries agencies has been to preserve the critical habitats upon which the long-term productivity of the fisheries depends. For penaeid prawns this has meant that seagrass (tiger prawns), and mangroves (banana prawns) have achieved special status to fishers, fisheries biologists, managers and legislators. Is this justified? Is this the appropriate management strategy to preserve critical fisheries habitat? We examine these questions using two case studies: cyclones, seagrasses and tiger prawns in the Gulf of Carpentaria and king prawns in the Peel-Harvey estuarine system in Western Australia.

It is clear that a greater understanding of the key processes operating in the coastal zone is a critical requirement for fisheries management. It is not enough to just map, monitor and maintain subsets of these systems based on coarse distribution and abundance studies of prawn populations. With increasing pressure on the coastal zone from competing interest groups, fisheries managers need a greater understanding of the factors which determine the carrying capacity of nursery habitats for juvenile penaeid prawns, and the factors which limit the distribution of key fisheries habitats within coastal ecosystems. Fisheries scientists and managers need to develop the knowledge base and management procedures for the implementation of ecosystem management.

Introduction

The Australian Fishing Zone occupies an area 16% larger than the Australian continent. This is the third largest fishing zone in the world. Commercial fisheries in this zone were worth approximately A$1,200 million in 1991–92 (gross value), of which 80% was exported (Anon 1992). This was made up of a diverse array of single and multispecies fisheries, with over 150 commercial species. Most of these fisheries are regional or local, and most stocks are dependent upon near shore or coastal nursery habitats.

Recently fisheries management in Australia has shifted to emphasise the management of the resources within the principles of ecologically sustainable development. This has resulted in management to sustain fish stocks, maximise economic efficiency when harvesting those stocks, and a trend towards granting property rights to the fishers (Anon 1991). To achieve the goal of management to sustain fish stocks, a major focus of fisheries agencies has been to preserve the critical habitats upon which the long-term productivity of the fisheries depends.

For penaeid prawns this has meant that seagrasses (the critical nursery habitat for tiger prawns, Figure 1) and mangroves (the critical nursery habitat for banana prawns, Figure 2) have achieved special status to fishers, fisheries biologists, managers and legislators; whereas other key habitats have not received the same special status e.g. shallow sandy substrates (a critical nursery habitat for king prawns, Potter *et al.* 1991). Is this justified? Is this the appropriate management strategy to preserve critical fisheries habitat? We examine these questions using two case studies: cyclones, seagrasses and tiger prawns in the Gulf of Carpentaria and king prawns in the Peel-Harvey estuarine system in Western Australia.

Cyclones, seagrasses and prawns— Gulf of Carpentaria

The Gulf of Carpentaria is a large, rectangular (approx. 3.7×10^5 km^2), shallow (<70 m), tropical embayment between 11–17.5° S latitude and 136–142° E longitude (Rothlisberg and Jackson 1982). The area has marked seasonality in temperature, salinity, rainfall and wind regimes (Poiner *et al.* 1987). Rainfall is restricted to the north-western monsoon in summer (December to February) and there is a very dry period from May to October during the south-east trade winds (Poiner *et al.* 1987).

Commercial prawn fishing in the Gulf of Carpentaria began in the late 1960s and initially concentrated on the banana prawn (*Penaeus merguiensis* de Man) (Somers *et al.* 1987). Tiger prawns (*Penaeus esculentus* Haswell and *P. semisulcatus* de Haan) are now the most important component of the catch, with 3000 to 4000 tonnes caught each year, mostly in the Western Gulf (Somers *et al.* 1987). The juvenile stages of both species of tiger prawns are most commonly found in seagrass beds (Figure 1) (Staples *et al.* 1985). The seagrasses of the Gulf of Carpentaria were mapped in 1982, 1983 and 1984. There were approximately 906 km^2 of seagrass beds in the Gulf, fringing 671 km of coastline, and consisting of eleven different seagrass species (Poiner *et al.* 1987).

On average five cyclones occur on the Australian coastline each year, although the frequency and track vary from year to year. In 1985 cyclone Sandy approached the coast at the Sir Edward Pellew group of Islands. Unlike many cyclones it travelled parallel to the coast, and finally crossed north of the Roper River (Poiner *et al.* 1989). The western Gulf of Carpentaria, including the area affected by cyclone 'Sandy', had been surveyed by CSIRO in 1984, immediately prior to the cyclone. The distribution of seagrass beds, their species composition, density, morphology and biomass were recorded. Following the cyclone, the affected area and nearby 'control' areas unaffected by cyclone Sandy were surveyed annually from 1985 to 1990, and then again in 1992. In the last four trips the juvenile prawn communities in inshore areas were sampled in the affected and control areas.

In 1985, immediately after the cyclone, the inshore seagrass beds in the area affected by the cyclone had disappeared. Seagrass in the deeper offshore water had been severely disturbed, but still survived. In 1986, there was still no seagrass inshore and the deep water beds had also disappeared. In all, 183 km^2, or 18-20% of the seagrass in the Gulf of Carpentaria was removed by cyclone Sandy (Poiner *et al.* 1989). Recolonisation was first recorded in 1987, two years after the cyclone, when a few shallow inshore areas were sparsely covered with patches of *Halodule uninervis* (Poiner *et al.* 1989). By 1988 about 20% of the area affected by cyclone Sandy had been recolonised by seagrass. *Halodule uninervis* and *Halophila ovalis*, which are of little value as habitat for juvenile prawns, were the predominant species. However a few isolated seedlings of the more useful *Cymodocea serrulata* and *Syringodium isoetifolium* were also recorded. In 1989 and 1990 the areal extent of seagrass did not change significantly, but there was an increase in species diversity of seagrass in the area colonised by *C. serrulata* and *S. isoetifolium*. By 1990, an area approximately 40 km long, south of the Limmen Bight River, had still not been recolonised.

In the area affected by cyclone Sandy the most common juvenile prawns observed in 1989 and 1990 were mostly small non-commercial species, mainly belonging to the genus *Metapenaeus*. In contrast, commercially important tiger and endeavour prawns were found in the undamaged seagrass beds (Thorogood *et al.* 1990).

Log book data and landing statistics from 1980 to 1991 were analysed to determine whether there was a decline in the catch of tiger prawns after the destruction of the seagrass beds. The catch of tiger prawns in the South western Gulf of Carpentaria fluctuates widely from year to year. From 1980 to 1984 the average annual catch of tiger prawns in both the affected and unaffected areas was about 250 tonnes (Thorogood *et al.* 1990). Since cyclone Sandy in 1985 the annual catch in the unaffected area has ranged from 100 to 350 tonnes, with an average of 200 tonnes, while the total catch in the affected area declined to about 40 tonnes in 1988 and 1989 (Thorogood *et al.* 1990), and in 1991 was 87 tonnes (Figure 3). That is, the loss through the cyclone, of seagrass as a habitat for juvenile prawns, may have resulted in a decrease in the commercial prawn catch in the fishery immediately offshore of the affected area of up to 80%, or 160 tonnes.

The total catch of tiger prawns in the whole of the Gulf of Carpentaria also fluctuates widely from year to year. The annual catch from 1980 to 1991 in the Gulf was 3,848 tonnes. So the conjectured loss of 160 tonnes of tiger prawns due to the effects of the cyclone is approximately 4% of the annual average catch for the Gulf, despite an 18–20% loss of seagrass in the Gulf.

Thus the severe effect on the juvenile prawn habitat and commercial prawn fishery is localised, and was not reflected in the total commercial catch. This begs the question: How much seagrass habitat can we lose before there is a severe effect on the fishery? Juvenile tiger prawn abundance can vary greatly between seagrass communities of different types and different tiger prawn species appear to prefer different seagrass habitat types, which probably explains the relatively small impact of Cyclone Sandy on the annual average catch for the Gulf.

Peel-Harvey estuarine system

The Peel-Harvey Estuary (lat $32°\ 35'$ S, long $115°\ 45'$E) is located 80 km south of Perth and is the largest estuarine system in south-western Australia, covering a surface area of about 130 km^2. It consists of two shallow (mostly 2 m deep) inter-connected basins (Peel Inlet and Harvey Estuary) and a short, narrow Entrance Channel linking the system to the sea (McComb *et al.* 1981). This estuary undergoes large fluctuations in salinity during the winter and spring months when approximately 90% of the annual rainfall is recorded in this region, and much of the Estuary can become hypersaline (up to 50 ppt) during late summer, early autumn (McComb *et al.* 1981; Loneragan *et al.* 1986). The Peel-Harvey system supports important commercial and recreational fisheries for a variety of species of fish and crustaceans (Potter *et al.* 1983; 1991).

Since the late 1960s, the Peel-Harvey Estuary has shown increasing signs of eutrophication due to the high levels of nutrients in the run-off derived from the agricultural lands in the catchment of the system. Initially there was a very high biomass of macroalgae, particularly the goat weed *Cladophora montagneana*, in the 1970s. However, in the 1980s and early 1990s the biomass of goat weed in the system declined dramatically and there have been virtually annual blooms of the cyanobacteria *Nodularia spumigena* during the summer months (McComb *et al.* 1981; Lukatelich and McComb 1986).

The estimated total annual biomass of *Cladophora* reached a peak of 26 000 t in 1979 and has been lower than 2 000 t throughout most of the 1980s and early 1990s (Lukatelich and McComb, unpublished data). This species covers the bottom and in the 1970s formed very dense, deep beds which smothered the substrate in large areas of the Peel-Harvey Estuary. The large banks of this species and the breakdown of macroalgae in the shallows have caused odour problems for residents and tourists in the region. A macroalgae harvesting program has been undertaken to remove some of the extensive beds of macroalgae in an attempt to alleviate some of this problem.

Western king prawns, *Penaeus latisulcatus*

Adults of the western king prawn *Penaeus latisulcatus*, are found on sandier substrates in the Gulf of Carpentaria and Exmouth Gulf, Western Australia than tiger prawns (Penn and Stalker 1979; Somers 1987; Dall *et al.* 1990). This species is much sought after by both recreational and commercial fishers in the Peel Harvey Estuary who catch the large juveniles as they emigrate from the estuary on the ebb tides at night, mainly between March and July of each year (Potter *et al.* 1991). Commercial catch data show that catches in the 1960s were much higher than those in the 1970s when *Cladophora* reached very high biomasses in the system (Figure 4) (Potter *et al.* 1991). Following the decline in biomass of *Cladophora*, the commercial catches increased greatly in the 1980s (Figure 4) (Potter *et al.* 1991).

The marked decline in catches is probably due to the loss of extensive areas of the sandy substrate, the required nursery habitat of this species (Penn and Stalker 1979; Dall *et al.* 1990; Potter *et al.* 1991). Although the biomass of other species of macroalgae has been high in the estuary after 1979, it has not reached the same levels as that reached by *Cladophora*. Moreover, these other species of macroalgae do not smother the substrate or form as dense and extensive beds as *Cladophora*. It would appear that although the system is still highly eutrophic in the 1980s and early 1990s, the recovery of the sandy nursery grounds of the western king prawn in the Peel-Harvey Estuary has led to a recovery of the commercial catches of this species.

Discussion and conclusions

In the relatively pristine Gulf of Carpentaria, the site of Australias major tiger and banana prawn fisheries, a natural decline of around 20% (183 km^2) of prime seagrass habitat resulted in a 4% (160 t) decline in the total catch of the fishery.

A simplistic analysis of the data would suggest the fishery can be sustained despite significant declines in coastal seagrass habitats. In the highly eutrophic Peel-Harvey Estuary, loss of sand substrate through smothering by a macroalga (*Cladophora montagneana*) in the 1970s, led to a marked decline in catches of western king prawn (*Penaeus latisculatus*). In the 1980s the system is still highly eutrophic but with the decline of the macroalgae and partial recovery of the sand habitat, catches of king prawns have recovered. It is clear from both of these studies we do not understand in detail the relationship between prawns and their nursery habitats or the factors that limit the distribution of habitats themselves.

Without suitable nursery areas, there would be no prawn fisheries. But to protect nursery areas and hence the long-term productivity of a fishery, a fishery manager has to know what it is that needs protecting. What exactly are the nursery habitats and what is it that makes some habitats more suitable than others? Juvenile tiger prawns are most abundant on seagrass beds, and juvenile banana prawns are most abundant in mangrove-lined estuaries. Indeed, in the Northern Prawn Fishery, there is good agreement between the distribution of the main tiger prawn fishing grounds and the distribution of coastal seagrasses, and between banana prawns and adjacent mangrove-lined estuaries (Staples *et al.* 1985). However, juvenile tiger prawn abundance can vary greatly between seagrass communities of different types and different tiger prawn species appear to prefer different seagrass habitat types. In the case of banana prawns, no one yet knows whether different types of mangrove communities support different population densities but, based on the strong regional variability in commercial catches, we suspect that this may certainly be the case. Just as importantly, no one yet knows what limits the distribution of the nursery habitats themselves. Clearly, it is important for fishery managers to know what the most suitable nursery habitats are so that they know *what* to protect; but it is just as important to know what factors make habitats suitable so that managers might know *how* to protect them. Where these habitats have been impacted and/or are limiting the productivity of the fishery, obviously it would also be advantageous to know what factors limit the growth and colonisation of nursery habitat vegetation and hence the distribution of habitats themselves.

It is clear that a greater understanding of the key processes operating in the coastal zone is a critical requirement for fisheries management. It is not enough to just map, monitor and maintain subsets of these systems based on coarse distribution and abundance studies of prawn populations. With increasing pressure on the coastal zone from competing interest groups, fisheries managers need a greater understanding of the factors which determine the carrying capacity of nursery habitats for juvenile penaeid prawns, and the factors which limit the distribution of the key fisheries habitats within coastal ecosystems. Fisheries scientists and managers need to develop the knowledge base and management procedures for the implementation of ecosystem management. We need to broaden our focus from the commercial industry and the populations of the target species and their critical habitats to a better understanding of the ecosystems within which the target species and their critical habitats are located (Anon 1991).

References

Anonymous (1991). Ecologically sustainable development working groups. Final Report—Fisheries. Australian Government Publishing Service, Canberra, 202pp.

Anonymous (1992). Background Fisheries Statistics. Department of Primary Industries and Energy, Canberra. Issue No. 6, 1992.

Dall, W., B.J. Hill, P.C. Rothlisberg and D.J. Staples (1990). The Biology of Penaeidae. *Advances in Marine Biology* **27**.

Loneragan, N.R., I.C. Potter, R.C.J Lenanton and N. Caputi (1986). Spatial and seasonal differences in the fish fauna of the shallows in a large Australian estuary. *Marine Biology* **92**, 575–586.

Lukatelich, R.J. and A.J. McComb (1986). Nutrient levels and the development of diatom and blue-green algal blooms in a shallow Australian estuary. *Journal of Plankton Research* **8**, 597–618.

McComb, A.J., M.L. Cambridge, H. Kirkman and J. Kuo (1981). Biology of seagrasses. In *The Biology of Australian Plants*. (Eds: J.S. Pate and A.J. McComb), 258–93. (The University of Western Australia Press, Nedlands).

Penn, J.W. and R.W. Stalker (1979). A daylight sampling net for juvenile penaeid prawns. *Australian Journal of Marine and Freshwater Research* **26**, 287–291.

Poiner, I.R., D. J.Staples and R. Kenyon (1987). Seagrass communities of the Gulf of Carpentaria, Australia. Australian Journal of Marine and Freshwater Research **38**, 121–31.

Poiner, I.R., D.I. Walker and R.G. Coles (1989). Regional studies—seagrasses of tropical Australia. In: *Biology of Seagrasses—An Australian Perspective* (Eds: Larkum, A.W.D., A.J. McComb and S.A. Shepherd), 279–303. (Elsevier, Amsterdam).

Potter, I.C, R.J.G Manning and N.R. Loneragan (1991). Size, movements, distribution and gonadal stage of the western king prawn (*Penaeus latisulcatus*) in a temperate estuary and local marine waters. *Journal of Zoology, London* **223**, 419–445.

Potter, I.C., N.R. Loneragan, R.C.J Lenanton, P.J. Chrystal and C.J. Grant (1983). Abundance, distribution and age structure of fish populations in a western Australian estuary. *Journal of Zoology, London* **200**, 21–50.

Rothlisberg, P. C. and C.J. Jackson (1982). Temporal and spatial variation of plankton abundance in the Gulf of Carpentaria, Australia 1975–1977. *Journal of Plankton Research* **4**, 19-40.

Somers, I.F. (1987). Sediment type as a factor in the distribution of the commercial prawn species of the western Gulf of Carpentaria, Australia. *Australian Journal of Marine and Freshwater Research* **38**, 133–149.

Somers, I.F., P.J. Crocos and B.J. Hill (1987). Distribution and abundance of the Tiger Prawns *Penaeus esculentus* and *P. semisulcatus* in the north-western Gulf of Carpentaria, Australia. *Australian Journal of Marine and Freshwater Research* **38**, 63–78.

Staples, D.J., D.J. Vance and D. S. Heales (1985). Habitat requirements of juvenile penaeid prawns and their relationship to offshore fisheries. In P.C. Rothlisberg, B.J. Hill and D.J. Staples (Editors) Second Australian National Prawn Seminar NPS2, Cleveland, Australia.

Thorogood, C.A., I.R. Poiner, I.F. Somers and D. Staples (1990). Seagrass and cyclones in the western Gulf of Carpentaria, 1989 update. CSIRO Marine Laboratories Information Sheet No. 7, January 1990.

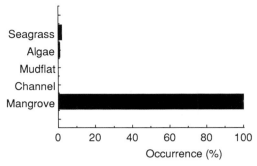

Figure 2. Percentage distribution of the catch of banana prawns (*Penaeus merguiensis*) in each of five habitats in the Embley River estuary, Gulf of Carpentaria.

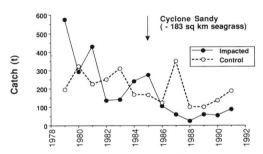

Figure 3. Catch of tiger prawns before and after cyclone Sandy in areas affected (impacted) and areas unaffected (control) by the cyclone.

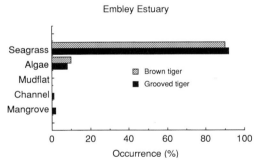

Figure 1. Percentage distribution of the catch of brown (*Penaeus esculentus*) and grooved (*P. semisulcatus*) tiger prawns in each of five habitats in the Embley River estuary, Gulf of Carpentaria.

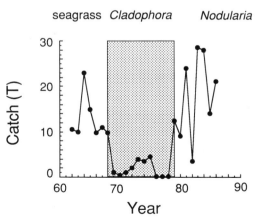

Figure 4. Catch of western king prawns (*Penaeus latisulcatus*) from the Peel-Harvey estuary before any algal blooms, during the blooms of the macroalgae *Cladophora montagneana* and during the summer blooms of the cyanobacteria *Nodularia spumigena* (redrawn from Potter *et al.* 1991).

REHABILITATION, MITIGATION AND RESTORATION OF FISH HABITAT IN REGULATED RIVERS

S. Swales

NSW Fisheries, Fisheries Research Institute
P.O. Box 21
Cronulla NSW 2230

Abstract

The principal factors implicated in the continuing decline of inland fisheries throughout the world are generally considered to be the reduction in quality and quantity of fish habitat conditions through the effects of land and water resource development works. In particular, river regulation through channelisation, impoundment, abstraction and other river engineering and management works often has major adverse effects on fish habitat conditions. However, a wide variety of techniques exists for restoring suitable habitat conditions in regulated rivers and improving fish abundance and diversity. For example, instream habitat improvement devices such as current deflectors and artificial cover devices can be used to re-create lost habitat features, such as the pool-riffle pattern, while artificial fishways improve habitat accessibility in impounded rivers. The provision of suitable environmental flows in regulated rivers is an essential requirement for sustaining fish habitat conditions. The use of such techniques in the rehabilitation of inland fisheries is illustrated through case-studies in the U.K., Canada and Australia.

Perturbations affecting the status of inland fisheries

Inland fisheries throughout the world are being increasingly impacted by a wide range of human activities which reduce the quality and quantity of available fish habitat and consequently adversely affect fish survival and abundance (Alabaster 1985). The major perturbations to the environment responsible for most loss and degradation of fish habitat arise from the effects of land and water resource development and river management, particularly works such as river impoundment for water supply and hydropower generation (Petts 1984). In addition, fish habitat conditions may be seriously affected by river channelisation for land drainage improvement and flood alleviation, water abstraction for crop irrigation, wetland drainage and removal of aquatic, riparian and floodplain vegetation as part of river and floodplain management (Brookes 1988). These perturbations may affect physical, chemical and biological aspects of fish habitat. In addition to such physical perturbations, inland fisheries are also affected by a wide range of other chemical and biological perturbations, including water pollution, over-exploitation and introduction of non-native species (Cowx in prep.).

Major physical perturbations to the river environment associated with impoundment and channelisation often affect both the quality and quantity of fish habitat. The construction of dams and weirs may have a wide range of effects on fish habitat, both upstream and downstream of impoundments (Petts 1984). The migratory movements of fish populations in both upstream and downstream directions are usually hindered or prevented by channel impoundment. River flow regulation by the impoundment has a variety of effects on downstream channel morphology, hydrology and water quality which may modify fish habitat conditions (Brooker 1981). River channelisation by channel dredging, widening or straightening may lead to an overall loss of available fish habitat and may severely degrade the remaining habitat. Such fundamental habitat features as channel meandering and the pool-riffle pattern may be severely modified by river channel works activities which usually lead to major reductions in overall habitat diversity (Swales 1982a).

River regulation activities such as these often produce major changes in the ecology of the aquatic environment, and fish populations may show marked declines in abundance and diversity through changes to habitat conditions, particularly reductions in the quantity and quality of physical habitat features and modifications to the natural flow regime (Petts 1984; Swales 1982a).

Fish habitat rehabilitation in regulated rivers

A wide variety of rehabilitation techniques exists for mitigating the impacts of human perturbations to the aquatic environment and fish habitat. Within the fields of river and fishery management there is a range of approaches available to improve the physical, chemical and biological aspects of fish habitat (Gore 1985). Possible measures to improve chemical or biological aspects of fish habitat may include pollution control, stocking or biomanipulation. Physical habitat restoration measures may include practices to maintain instream flows, riparian and aquatic vegetation, habitat diversity, macro-habitat features such as the pool-riffle pattern, and also preferred micro-habitat conditions (Wesche 1985). Habitat requirements of native fish show considerable variation between species and seasons, and also between age groups, which must be considered in designing habitat restoration programmes.

In the field of habitat restoration the terms mitigation, restoration, rehabilitation and amelioration are often used interchangeably and apparently arbitrarily when discussing methods for restoring fish habitat in regulated rivers. However, different approaches are available for improving fish habitat, depending on the extent of habitat degradation and whether the measures are proactive or reactive. Wherever possible it is preferable to practice impact mitigation in which the adverse effects of the perturbation are minimised using a proactive approach (analogous to 'preventative' medicine). This approach is receiving increasing attention as the preferred option in river conservation and management (see Boon *et al.* 1992). However, there are still many situations where a reactive approach to river conservation is the only option to sustain fish habitat in regulated rivers. These techniques will be considered briefly in the following discussion and illustrated with some overseas case studies.

Habitat restoration involves repairing damage caused by past activities and developments. Environmental compensation is the creation of freshwater habitats in order to compensate for anticipated adverse environmental effects of proposed developments. Habitat restoration can be either passive or active and proactive or reactive, while habitat compensation is generally active and proactive (Cairns 1990).

It is important to define the aim of habitat restoration in regulated rivers since it may not be practicable to aim to restore a stream or river to its pristine condition prior to the disturbance (Cairns 1990). A more practicable aim may be to minimise the loss and degradation of fish habitat in order to maintain a reasonably diverse and productive aquatic community. Alternatively, if the area affected contains an important recreational fishery it may be more important to aim to maximise fish density and biomass.

Habitat restoration techniques

Physical habitat improvement methodology

Fish habitat in streams and rivers lost or degraded by river channel works does recover in time as natural hydrologic processes cause the channel to readjust to the new conditions, and recovery processes restore and re-create lost habitat features. However, the natural recovery process is necessarily a long-term process and fish habitat in channelised rivers may take many years to recover to a level similar to the pre-impact conditions (Swales 1982a). However, the natural recovery process can be accelerated considerably through the use of remedial artifical habitat improvement devices which can be used to improve and modify habitat features. Such active measures are generally preferable from an environmental perspective since the goal of habitat restoration is achieved in a shorter time than it would by relying solely on passive natural recovery processes. In severely degraded streams, natural morphological and biological recovery may be a very slow process, typically requiring 10-100 years (Swales 1982a).

Structures such as low dams, current deflectors and artificial cover devices have been used for many years to improve habitat conditions for recreationally important species of trout and salmon in the streams and rivers of North America and Europe (Wesche 1985). However, these devices are now receiving increasing use and attention as a means of restoring lost or degraded fish habitat in regulated rivers (see review in Swales 1989). An increasing number of studies have documented the successful use of these devices in restoring fish habitat in rivers which have been channelised, impounded or otherwise impacted by river channel works activities. As a result, it has often been possible to partially restore the diversity and abundance of fish populations in impacted areas of stream or river (e.g. Swales and O'Hara 1983).

Artificial fishways

In addition to habitat improvement techniques, there is also a wide range of techniques available to improve the accessibility of suitable habitat areas. The construction of dams and weirs often considerably limits the availability of suitable habitat areas. However, there is now a well developed field concerned with the design and construction of artificial fishways in dams and weirs which allow the upstream and downstream movements of fish. Most fishways take the form of artificial 'fish-ladders' or other fish-passes which provide a free passageway between the lower and upper areas of the impoundment. Most fish-passes have been constructed for recreationally and commercially important fishes, particularly salmonid species of salmon and trout, in the impounded rivers of the northern hemisphere. In the Murray-Darling River in south-east Australia, however, fish-passes installed in several weirs to improve the passage of native fish species have been shown to be effective in allowing fish movements in the river (Mallen-Cooper 1989).

Environmental stream flows

The impoundment of streams and rivers using dams and weirs, together with water abstraction for crop irrigation, has the effect of regulating the natural river flow to produce a hydraulic flow regime which is considerably removed from the natural flow regime before river regulation (Petts 1984). An important aspect of the restoration of fish habitat conditions in regulated rivers is the assessment of suitable instream flows for fish and other biota. Instream flows provided for environmental reasons, designed to enhance or maintain the habitat for riparian and aquatic life, are often referred to as environmental flows (Gordon *et al.* 1992). A wide range of techniques have been developed to assess the suitability of stream flows for the survival of fish and other biota (see reviews in Kinhill 1988; Gordon *et al.* 1992).

The available techniques fall into three major categories.

(1) Historical discharge methods
These are based largely on historical flow records and use a fixed proportion of flow to provide environmental stream flow recommendations. They are also referred to as 'rule of thumb' methods since they do not involve the collection of field data, e.g. Tennant or Montana method (Tennant 1976).

(2) Transect or hydraulic rating methods
These involve the collection of field data at one or more transects in a stream reach and the development of relationships between discharge and other physical habitat variables (e.g. Stalnaker 1980).

(3) Instream habitat simulation methods
These consider not only how physical habitat changes with streamflow but combine this information with the habitat preferences of a given species to determine the amount of habitat available over a range of stream flows e.g. Instream Flow Incremental Methodology (IFIM) (Bovee 1982).

Case studies

a. Habitat restoration in a channelised river

Many lowland rivers in the United Kingdom are subject to periodic river engineering and management works aimed at increasing channel hydraulic capacity and modifying flow conditions to improve land drainage and alleviate flooding. River channel works such as channel realignment, bank regrading and vegetation clearance, dredging of the river bed and clearing of aquatic weeds and other instream material are frequently undertaken over large sections of lowland river by river management authorities. Such works may have severe effects on the quality and quantity of fish habitat and on river ecosystem processes (Swales 1982a). As a result, channelised rivers generally experience major declines in fish abundance and diversity as a response to the loss of major habitat features such as the pool-riffle pattern and the overall decline in habitat diversity (Swales 1982a; 1989).

Although habitat conditions in channelised rivers will improve in time through river recovery processes, studies have shown that natural recovery is a long-term process. As part of an experimental study, attempts were made to accelerate the natural recovery process in a channelised river in north-west England using artificial habitat improvement devices. The responses of the fish community to the habitat changes associated with the installation of these devices was assessed. Habitat improvement devices in the form of low weirs, current deflectors and artificial cover structures were installed in a channelised area of river and were found to be effective in improving fish habitat conditions (Swales 1982b). The low dams and current deflectors were found to be effective in re-creating habitat features such as the pool-riffle pattern, while the cover structures simulated areas of shelter normally associated with river banks.

Fish populations in the study section were monitored before and after the installation of the improvement devices and the population density and biomass of the main fish species were estimated by electrofishing. Following habitat improvement, the population densities of dace and chub, the two main species present, increased by 75% and 37% respectively, while population biomass increased by 31% and 25% (Swales and O'Hara 1983). Distribution mapping studies revealed considerable fish relocation following habitat improvement, with large concentrations of fish being recorded in the vicinity of the improvement structures (Figure 1). It was concluded that the improvement programme was successful in partially mitigating the adverse effects on the fishery of previous land drainage works.

b. Restoration of off-channel habitats

The streams and rivers of British Columbia, Canada, are major producers of anadromous species of salmonid such as rainbow trout (*Oncorhynchus mykiss*), coho salmon (*O.kisutch*) and chinook salmon (*O.tshawytscha*). Such species typically rear in freshwaters for several years before migrating to the ocean for adult maturation and development. Recent studies in the rivers of British Columbia have shown that different species of juvenile salmonid show considerable variation in their habitat preferences and utilisation. In addition, salmonid species show major seasonal differences in habitat preferences. For example, in summer, juvenile coho salmon generally prefer pool habitats with abundant cover, but in autumn and early winter, populations migrate from their summer rearing areas into tributaries, back-channels, sloughs, ponds and lakes in which they overwinter (Swales *et al.* 1986; 1988).

In the Coldwater River, a tributary of the Fraser River in interior British Columbia, juvenile coho salmon and other salmonid species were found to utilise off-channel habitats, particularly shallow floodplain ponds, as their preferred overwintering areas (Swales *et al.* 1986; Swales and Levings 1989). The ponds were found to contain high population densities of overwintering coho salmon (up to 4,000 per hectare) and growth in the ponds appeared to be faster than in main channel habitats. Habitat conditions in the ponds were less extreme than in the main stream channel. Off-channel ponds appeared to play a valuable role in the life cycle of coho salmon and other juvenile salmonids (Swales and Levings 1989).

As a result of road construction works along the valley of the Coldwater River, the stream channel was diverted and it was found necessary to drain several off-channel ponds. As a consequence of the policy of 'no net loss' of fish habitat operated by the Department of Fisheries and Oceans in Canada it was required that equivalent habitat be created by the developers to compensate for the loss of stream channel and pond rearing habitat. As a result, several artificial off-channel ponds were constructed and connected to the realigned stream channel to provide an equivalent area of rearing habitat for juvenile salmonids. Monitoring of the ponds shortly after construction showed that juvenile salmonid species were using the ponds as overwintering habitats and that the ponds were successful in partially compensating for the lost natural habitats (Swales *et al.* 1986).

c. Environmental streamflow assessment

The flow of many streams and rivers in the Murray-Darling river basin in south-east Australia is regulated by dams or weirs constructed for water supply, navigation or hydro-power generation. Approximately 75% of all the irrigated land in Australia is contained within the Murray-Darling basin and numerous storage impoundments have been constructed in the headwaters of eastern tributaries of the Darling River. Water supply for irrigation has resulted in major changes to the hydrologic flow regime of most of the rivers in the basin. Flow regulation is thought to be one of the principal factors implicated in recent declines in native fish populations in the Murray-Darling basin (Cadwallader 1986; Lloyd *et al.* 1991).

A wide range of methodologies exists for the assessment of the instream flow requirements of fish and other biota in regulated rivers (Gordon *et al.* 1992). However, most of these have been developed and tested overseas and may not be suitable for use in Australian rivers, where the streamflow regime is generally more unpredictable and variable than in most other countries. There is a need for the development of a reliable and simple method for recommending minimum instream flows to protect aquatic life in Australian streams and rivers which is inexpensive, easy to perform and requires little or no field investigation (Orth and Leonard 1990).

In New South Wales, the Department of Fisheries and the Department of Water Resources are currently investigating the use of an alternative approach based on the use of 'expert-panels' to assess streamflow suitability for fish populations and river ecosystem processes. In this approach, expert-panels are set up consisting of specialists in the fields of fish biology, river invertebrate ecology and fluvial geomorphology. The flow regime below storages is experimentally manipulated and a range of discharges released from the storage. The expert-panel is then asked to score the suitability of each flow release as an environmental flow in which native fish are the primary indicators of environmental quality. Preliminary results of the study suggest that the expert-panel approach potentially has a valid and important role to play in environmental streamflow assessments (Swales *et al.* in prep).

In general there was a consistent trend at all storages for the expert-panels to prefer the lowest discharge releases as summer flows, the highest discharge releases as winter flows, and intermediate discharges as spring and autumn flows. The expert-panel approach, despite some limitations and drawbacks, is thought to be potentially a useful tool in instream flow studies for assessing suitable environmental flows in regulated rivers.

Conclusions

Although river management and engineering works such as channelisation and impoundment can have wide-ranging adverse effects on inland fisheries through reductions in the quality and quantity of fish habitat, it is nonetheless often possible to restore and rehabilitate the river environment to mitigate the impacts of river channel works on fish populations. Wherever possible, however, it is still generally preferable to be proactive rather than reactive and to implement preventative measures to minimise the adverse effects of river modifications on fish habitat conditions. In the final analysis, fisheries in regulated rivers can only be sustained by taking measures which sustain fish habitat.

References

Alabaster, J.E.S. (Ed.) (1985). *Habitat modification and freshwater fisheries.* Butterworths, London.

Boon, P.J., P. Calow and G.E. Petts (1992). *River Conservation and Management.* Wiley & Sons, Brisbane.

Bovee, K.D. (1982). A Guide to Stream Habitat Analysis using the Instream Flow Incremental Methodology. Instream Flow Inf. Pap. 12, FWS/OBS-82/26, Office of Biological Service, Fish and Wildlife Service, U.S. Department of the Interior, Washington D.C.

Brooker, M.P. (1981). The impact of impoundments on the downstream fisheries and general ecology of rivers. pp 91–152 In, Advances in Applied Biology, Vol. 6. (ed. T.H. Coaker), Academic Press, New York.

Brookes, A. (1988). Channelised Rivers: *Perspectives for Environmental Management.* John Wiley, Chichester.

Cadwallader, P.L. (1986). Fish of the Murray-Darling system. In, *The Ecology of River Systems* (Eds. B.R. Davies & K.F. Walker). The Hague, Junk.

Cairns, J. (1990). Lack of theoretical basis for predicting rate and pathways of recovery. *Environmental Management* **14**, 517–526.

Cowx, I.A. (in prep). Rehabilitation of Inland Fisheries. Proceedings of the Workshop and International Symposium on the Rehabilitation of Inland Fisheries, University of Hull, April 1992.

Gordon, N.D., T.A. McMahon and B.L. Finlayson (1992). *Stream Hydrology: An Introduction for Ecologists.* Wiley & Sons, Chichester.

Gore, J.A. (1985). *The Restoration of Rivers and Streams.* Butterworths, London.

Kinhill (1988). Techniques for determining environmental water requirements—review. *Technical Report series,* Report No. 40, Kinhill Engineers Pty. Ltd, a report to the Department of Water Resources, Victoria.

Lloyd, L., J. Puckridge and K.F. Walker (1991). The significance of fish populations in the Murray-Darling system and their requirements for survival. pp. 86–99 In, *Conservation in the Management of the River Murray system—making conservation count.* Proceedings of the Third Fenner Conference on the Environment, Canberra, September, 1989.

Mallen-Cooper, M. (1988). Fish passage in the Murray-Darling basin. pp. 123–137 In, Proceedings of the Workshop on Native Fish Management, Canberra, June 1988. Murray-Darling basin Commission, Canberra.

Orth, D.J. and P.M. Leonard (1990). Comparison of discharge methods and habitat optimisation for recommending instream flows to protect fish habitat. *Regulated Rivers Research and Management* **5**, 129–138.

Petts, G.E. (1984). *Impounded Rivers: Perspectives for Ecological Management*, John Wiley & Sons, Chichester.

Stalnaker, C.B. (1980). The use of habitat structure preferenda for establishing flow regimes necessary for maintenance of fish habitat. pp. 321–337 In, *The Ecology of Regulated Streams* (Eds. J.V. Ward and J.A. Stanford), Plenum Press, New York.

Swales, S. (1982a). Environmental effects of river channel works used in land drainage improvement. *Journal of Environmental Management* **14**, 103–126.

Swales, S. (1982b). Notes on the construction, installation and environmental effects of habitat improvement structures in a small lowland river in Shropshire. *Fisheries Management* **13**, 1–10.

Swales, S. (1988). Fish populations of a small lowland channelised river in England subject to long-term river maintenance and management works. *Regulated Rivers Research and Management* **2**, 493-506.

Swales, S. (1989). The use of instream habitat improvement methodology in mitigating the adverse effects of river regulation on fisheries. pp. 185–209 In, *Alternatives in Regulated Rivers Management* (J.A. Gore & G.E. Petts, eds.) CRC Press, Boca Raton, Florida.

Swales, S. and C.D. Levings (1989). Role of off-channel ponds in the life-cycle of coho-salmon (*Oncorhynchus kisutch*) and other juvenile salmonids in the Coldwater River, British Columbia. *Canadian Journal of Fisheries and Aquatic Sciences* **46**, 232–242.

Swales, S. and K. O'Hara, (1980). Instream habitat improvement devices and their use in freshwater fisheries management. *Journal of Environmental Management* **10**, 167–179.

Swales, S. and K. O'Hara, (1983). A short-term study of the effects of a habitat improvement programme on the distribution and abundance of fish stocks in a small lowland river in Shropshire. *Fisheries Management* **14**, 135–145.

Swales, S., R.B. Lauzier and C.D. Levings (1986). Winter habitat preferences of juvenile salmonids in two interior rivers in British Columbia. *Canadian Journal of Zoology* **64**, 1506–1514.

Swales, S., F. Caron, J.R. Irvine and C.D. Levings (1988). Overwintering habitats of coho salmon (*Oncorhynchus kisutch*) and other juvenile salmonids in the Keogh River system, British Columbia. *Canadian Journal of Zoology* **66**, 254–261.

Swales, S., H. Cross and J.H. Harris (in prep). The use of an 'expert-panel' approach in assessing environmental flows in regulated rivers in south-east Australia.

Tennant, D.L. (1976). Instream flow regimens for fish, wildlife, recreation and related environmental resources. *Fisheries* **1**, 6–10.

Wesche, T.A. (1985). Stream channel modifications and reclamation structures to enhance fish habitat. pp. 103–159 In, *The Restoration of Rivers and Streams* (ed. J.A. Gore), Butterworth, Boston.

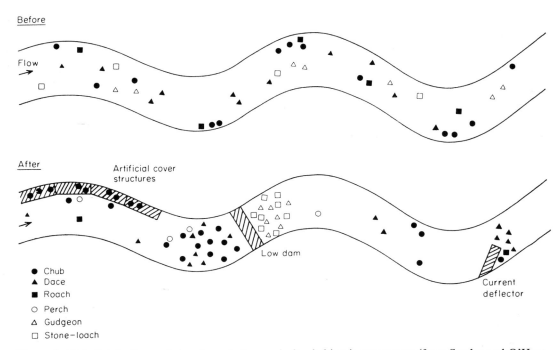

Figure 1. Fish distribution and abundance before and after habitat improvement (from Swales and O'Hara 1983).

A HABITAT FIT FOR FISH—
AN AIM OF BIOMANIPULATION?

M.T. Bales

Environmental Studies Unit, Department of Water Resources
10 Valentine Avenue
Parramatta NSW 2124

Abstract

A recurring problem has been observed world-wide whereby eutrophication has caused an increase in algal biomass and has frequently resulted in the occurrence of blue-green algal blooms. Biomanipulation, the 'top-down' manipulation of food chains as a means of reducing algal biomass, has been the subject of considerable research effort in the northern hemisphere and has been used in a number of cases as a means of managing and restoring eutrophic lakes. Its track record has been variable, a number of successes being countered by examples where the manipulation has failed to have the desired result. Some possible reasons for these failures, the complexity of the systems, the resistance of blue-green algae to grazing and the instability of the manipulated fish populations, are discussed. Particular attention is paid to the latter with the likely responses of fish populations following their manipulation being explored. In terms of the future use of biomanipulation in Australia it is suggested that there are four key questions: 1) how can the fish manipulation best be achieved? 2) what are the important buffering mechanisms ensuring the sustainability of the manipulation? 3) how can these buffering mechanisms be promoted and 4) have we the necessary information on fish biology, behaviour and ecology to determine how best to manipulate the system?. The importance of the future involvement of fisheries scientists in biomanipulation research and use is stressed.

Background to biomanipulation

Since the mid 1970's, when the term "biomanipulation" was used to describe the restructuring of the biological community as an approach to combating eutrophication (Shapiro *et al.* 1975), there has been considerable research effort throughout the northern hemisphere to assess and determine the effectiveness of the technique. With the problems caused by eutrophication increasing over recent decades, interest in biomanipulation has increased as other, more engineering type, solutions have proven to have limited success. This culminated in an international conference entitled "Biomanipulation Tool for Water Management" (Gulati *et al.* 1990) at which the current experimental and practical experience of biomanipulation was discussed, and the shortcomings in knowledge identified. The original concept of biomanipulation espoused by Shapiro *et al.* (1975) was the 'manipulation of the biota of lakes and their habitats to facilitate certain interactions and results which we as lake users consider beneficial'. The word biomanipulation is now normally used in a more restricted sense to refer to food web, or 'top-down', manipulation of a system (Shapiro 1990). In the majority of cases the reason for manipulating a system is to reduce algal biomass, in particular when blue-green algae are a problem. The basic concept behind all food-web manipulations is summarised in Figure 1. Its success depends upon the extent of predation on the zooplankton being reduced, ideally leading to an increase in the number of larger-sized individuals which are the most efficient grazers on phytoplankton and are thus best able to maintain a lower algal biomass. In the majority of cases the most significant predation pressure exerted on the zooplankton is due to fish. Manipulation of the fish population takes two main forms, removal of the zooplanktivorous fish resulting in a direct reduction in predation on the zooplankton, or enhancement of the piscivorous fish population as a means of reducing the zooplanktivorous fish. Examples exist where both approaches, or a combination of the two, have been used in an attempt to restore whole lakes. For example planktivorous fish were removed from Cockshoot Broad, U.K. in an attempt to reduce algal biomass, improve the underwater light climate and promote the growth of aquatic plants (B. Moss, personal communication). This approach has also been used at Lake Vaeng, Denmark (Søndergaard et al. 1990). More typically a combination of the two approaches has been used where the existing planktivorous populations are reduced as far as possible and the piscivorous fish populations are enhanced. Such an approach has been used as a restorative technique, for example for Lake Zwemlust (Van Donk et al. 1989; 1990a), Lake Bleiswijkse Zoom (Meijer et al. 1989) and Lake Breukeleveen (Van Donk et al. 1990b) in the Netherlands, Lake Frederiksborg Slotssø (Riemann et al. 1990) in Denmark, Lake Mosvatn (Sanni and Waervagen 1990) in Norway, and Lake Christina (Hanson and Butler 1990) in the U.S.A.

In Australia there has been very little work on biomanipulation. Geddes (1986) compared zooplankton communities in four farm dams, three of which were without fish and one stocked with redfin (*Perca fluviatilis*). He demonstrated differences in the species composition, abundance and size distribution of the zooplankton community which he attributed to the presence or absence of fish. Merrick and Ganf (1988) used enclosures to investigate the effects of zooplankton grazing on phytoplankton in Mount Bold Reservoir, and demonstrated some reduction in phytoplankton biomass in grazed enclosures. Lund (1991) used small enclosures to assess the impact on the foodweb of mosquito fish control but concluded that it led to no significant improvement in water quality. To this time there has been no attempt to use biomanipulation of the food chain as a management tool for the restoration of waterbodies in Australia.

Biomanipulation—success or failure?

A number of enclosure experiments have demonstrated an increase in larger zooplankton in the absence of planktivorous fish (e.g. Andersson et al. 1978) and a reduction in phytoplankton biomass in the presence of more or larger zooplankton (e.g. Schoenberg and Carlson 1984). These provide strong evidence that biomanipulation works on a small experimental scale, but can it be scaled up to the 'whole lake' scale?. Evidence from those lakes which have been manipulated is contradictory. In some instances the effects have been dramatic with examples of a desired increase in the numbers and/or size of the zooplankton, a decrease in the phytoplankton biomass, increased water clarity, increased macrophyte growth and an improvement in the amenity, recreational, aesthetic and drinking water quality. For example removal of planktivorous and benthivorous fish, and the addition of piscivorous fish to Lake Zwemlust, a small lake (1.5 ha, 1.5m deep) in the Netherlands resulted in a reduction in the chlorophyll *a* concentration from in excess of 250 µg l^{-1} to less than 5 µg l^{-1} and eliminated the appearance of *Microcystis* blooms. There was an increase in the number of larger zooplankton and an increase in the previously sparse aquatic vegetation (Van Donk et al. 1990a). On a larger scale at Lake Christina (1619 ha, 1.25m deep) in the U.S.A., removal of the fish population, and restocking with a more balanced fish population, resulted in a shift to larger zooplankton species, a reduction in chlorophyll *a* concentrations, an improvement in water transparency and an increase in submerged macrophytes (Hanson and Butler 1990).

In contrast many other attempts to biomanipulate lakes have failed to have the desired effect, or the initial improvements have been transient. For example in Lake Breukeleveen in the Netherlands (180 ha, 1.45m deep) removal of a substantial proportion (60-70%) of the planktivorous and benthivorous fish, the addition of 0+ piscivores, and seeding with large-sized daphnids failed to result in a reduction in the chlorophyll *a* concentration, improve the water transparency or increase the growth of aquatic plants (Van Donk *et al.* 1990b).

Why do biomanipulations fail?

The reasons for the difficulties encountered in making biomanipulations work are complex but perhaps fall into three categories. The basic food-web interactions are well understood, having been explored extensively in a large number of field enclosure and laboratory experiments, but the overall function of a lake or river system is complex and the end result of a manipulation is therefore difficult to predict. This may be particularly true in situations where there are significant populations of invertebrates which are predators on zooplankton and in turn are fed upon by fish. In such a case, removal of the fish population may result in an increase in these invertebrates, possibly increasing the overall predation pressure on the zooplankton, the reverse of the desired outcome. Secondly, the algae of greatest concern, the blue-greens, may be inedible and/or suppress the development of large-bodied zooplankton (Lampert 1981; Haney 1987). If so, reduction in the predation pressure exerted on the zooplankton by the fish is not going to feed through to a reduction in the phytoplankton biomass, and may even favour the development of blue-green algae. Finally, and of most interest in the context of this paper, the manipulated fish populations are unstable. Figure 2 summarises the typical response of fish populations following their manipulation. Where an attempt is made to remove the zooplanktivorous fish it is virtually impossible to succeed in removing 100% of the population. The remaining fish, with an abundance of food and spawning sites and in the absence of piscivorous fish are likely to achieve good breeding success, resulting in a large number of 0+ fish which will have a detrimental effect upon the zooplankton populations, undermining the aim of the manipulation. In the absence of a significant breeding success such a manipulation may succeed, a stable and desirable reduced fish population becoming established. If a 100% fish kill is achieved and the desired changes in the zooplankton and phytoplankton populations occur, the long term stability of the change depends upon the timing and nature of the re-establishment of the fish population. Of the three options, a total failure of fish to recolonise is unlikely, and although perhaps desirable in terms of the control of algal problems, is not desirable from a fisheries point of view. Almost certainly fish will recolonise a waterbody within a relatively short period, the key point being whether it reverts to something similar to the original undesirable structure, a new undesirable structure or a desirable fish population. In a more complex manipulation where the piscivorous fish populations are enhanced in addition to the decrease in the zooplanktivorous fish, there is again a range of possible outcomes. The environmental conditions may be unsuitable for the piscivores, resulting in a failure of them to survive, effectively the manipulation then being simply a removal of zooplanktivorous fish, with the potential problems discussed previously. Alternatively the piscivores may survive and effectively control the zooplanktivores, preventing their increase to the point where they undermine the success of the manipulation, or they may survive but fail to control the zooplanktivores.

Thus we see that fish are a key component determining the success or otherwise of biomanipulation. In much previous work on biomanipulation, fish have been considered a 'problem' to be eliminated in order to achieve success. If biomanipulation is to be used in Australia, there are two approaches we can take, either continuing to view the fish as a component of the system which must be removed, or alternatively considering fish to be an integral part of the system, and aiming to achieve the reduced phytoplankton populations in tandem with the creation of habitats suitable for the maintenance of desirable fish populations.

This poses four key questions:

1) How can the initial manipulation of the fish populations best be achieved?

2) What are the important buffering mechanisms which in the long term will allow the coexistence of sustainable and desirable fish, zooplankton, plants and phytoplankton?

3) How can we promote the buffering mechanisms identified in 2)?

4) Do we know enough about the habitat requirements of fish to determine how to manipulate the environment in order to encourage the desired species and minimise undesirable species?

The first is largely a practical question of how to manipulate the fish population. Physical methods of fish capture are well developed and readily usable but not suitable or feasible in all circumstances. Chemical control methods, while being less expensive and more effective, may run into public relations and legislative difficulties. More difficult to answer might be the determination of which piscivorous species are suitable for control of zooplanktivorous fish, and knowing what their environmental requirements are so that their populations can successfully be enhanced.

The second question is one of the role of buffering mechanisms. In the long term, in order to establish a degree of stability in the system, buffering mechanisms are required to prevent the new desirable fish populations from reverting to their pre-manipulation undesirable state. This may simply involve a change in the species present. For example a piscivore, previously absent or only present in small numbers, may act as a buffer to the redevelopment of troublesome populations of zooplanktivores. Another important buffering mechanism might be the presence of some kind of refugia for zooplankton, allowing their coexistence with zooplanktivorous fish. Different types of refugia were discussed by Shapiro (1990). They include, for example, low light, cool temperatures, low dissolved oxygen, predator inefficiency, macrophytes and physical refuges. In shallow lakes the presence of macrophytes probably represents the best form of refugia. Macrophytes have been shown to act as a refuge for larger *Daphnia*, decreasing predation by planktivorous fish (Timms and Moss 1984). It was suggested that the presence of macrophytes in the Norfolk Broads, U.K. allowed the maintenance of clear water and low phytoplankton biomass in the face of very high nutrient concentrations (Moss *et al.* 1985).

This leads to the third question. If buffering mechanisms are identified which are important for the long term success of a biomanipulation how can they actually be created in the field? For example if macrophytes are shown to be a useful buffering mechanism how can their growth be encouraged.

Finally if we are going to attempt to manipulate fish populations do we know enough about the habitat requirements of Australian fish to determine how to manipulate the environment in order to encourage the desired, and minimise the undesirable, species? Probably of greatest importance is an understanding of the requirements of a small number of piscivorous species, which would be suitable species to encourage, and of those zooplanktivorous species which can be identified as posing the greatest threat to zooplankton populations.

The future of biomanipulation in Australia

Over recent times we have observed an increase in the problems associated with eutrophication, particularly an increase in the occurrence of blue-green algal blooms. It has to be recognised that no single answer or approach exists which can solve all of these problems, but biomanipulation may represent one important weapon in overcoming them, either as a stand-alone technique, or in conjunction with other methods such as a reduction in nutrient concentrations. Fish are a central component of the biomanipulation process and it is important that fisheries scientists are actively involved in biomanipulation research to ensure that the existing knowledge and expertise on fish biology, behaviour and ecology is fully utilised and that the sustainability of fish populations, and the maintenance, creation or improvement of fish habitat can be integrated within the desired biomanipulation requirements, rather than fish again being considered purely as a problem to be overcome.

References

Andersson, G., H. Berggren, G. Cronberg and C. Gelin (1978). Effects of planktivorous and benthivorous fish on organisms and water chemistry in eutrophic lakes. *Hydrobiologia* **59**, 9–15.

Geddes, M.C. (1986). Understanding zooplankton communities in farm dams: the importance of predation. pp. 387-401 In *Limnology in Australia* (Eds. P. De Deckker and W.D. Williams) (CSIRO/Dr W. Junk Publishers).

Gulati, R.D., E.H.R.R. Lammens, M.L. Meijer and E. van Donk (1990). *Biomanipulation Tool for Water Management*. (Kluwer Academic Publishers), 628pp.

Haney, J.F. (1987). Field studies on zooplankton-cyanobacteria interactions. *New Zealand Journal of Marine and Freshwater Research* **21**, 467–475.

Hanson, M.A. and M.G. Butler (1990). Early responses of plankton and turbidity to biomanipulation in a shallow prairie lake. *Hydrobiologia* **200/201**, 317–327.

Lampert, W. (1981). Inhibitory and toxic effects of blue-green algae on *Daphnia*. *Internationale Revue der gesamten Hydrobiologie* **66**, 285–298.

Lund, M. (1991). An experimental study of foodweb manipulation and other strategies to control algal blooms at Lake Monger, Abstract-*Australian Society for Limnology 1991 Conference*.

Meijer, M.L., A.J.P. Raat and R.W. Doef (1989). Restoration by biomanipulation of Lake Bleiswijkse Zoom (The Netherlands): first results. *Hydrobiological Bulletin* **23**, 49–57.

Merrick, C.J. and G.G.Ganf (1988). Effects of zooplankton grazing on phytoplankton communities in Mount Bold Reservoir, South Australia, Using enclosures. *Australian Journal of Marine and Freshwater Research* **39**, 503–23.

Moss, B., H.R. Balls and K. Irvine (1985). Management of the consequences of eutrophication in lowland lakes in England-engineering and biological solutions. pp 180-185 In J.W. Lester and P.W.W. Kirk (Eds.), Proceedings International Phosphorus Conference. (SP Publishers, London).

Riemann, B., K. Christofferson, H.J. Jenson, J.P. Müller, C. Lindegaard and S. Bosselmann (1990). Ecological consequences of a manual reduction of roach and bream in a eutrophic, temperature lake. *Hydrobiologia* **200/201**, 241–250.

Sanni, S. and S.B. Wærvagon (1990). Oligotrophication as a result of planktivorous fish removal with rotenone in the small, eutrophic, Lake Mosvatn, Norway. *Hydrobiologia* **200/201**, 263–274.

Shapiro, J (1990). Biomanipulation: the next phase- making it stable. *Hydrobiologia* **200/201**, 13–27.

Shapiro, J, V. Lamarra and M.Lynch (1975). Biomanipulation: an ecosystem approach to lake restoration. pp 85-96 In *Proceedings of a Symposium on Water Quality Management through Biological Control*.(Eds. P.L. Brezonik and J.L. Fox). University of Florida, Gainesville, U.S.A.

Schoenberg, S.A. and Carlson, R.E. (1984). Direct and indirect effects of zooplankton grazing on phytoplankton in a hypereutrophic lake. *Oikos* **42**, 291–302.

Søndergaard, M., E. Jeppesen, E. Mortensen, E. Dall, P. Kristensen and O. Sortkjaer (1990). Phytoplankton biomass reduction after planktivorous fish reduction in a shallow, eutrophic lake: a combined effect of reduced internal P-loading and increased zooplankton grazing. *Hydrobiologia* **200/201**, 229–240.

Timms, R.M. and B. Moss (1984). Prevention of growth of potentially dense phytoplankton populations by zooplankton grazing, in the presence of zooplanktivorous fish, in a shallow wetland ecosystem. *Limnology and Oceanography* **29(3)**, 472–486.

Van Donk, E., R.D. Gulati and M.P. Grimm (1989). Food-web manipulation in Lake Zwemlust: positive and negative effects during the first two years. *Hydrobiological Bulletin* **23**, 19–34.

Van Donk, E., M.P. Grimm, R.D. Gulati and J.P.G. Klein Breteler (1990a). Whole-lake food-web manipulation as a means to study community interactions in a small ecosystem. *Hydrobiologia* **200/201**, 275–289.

Van Donk, E., M.P. Grimm, R.D. Gulati, P.G.M. Heuts, W.A. de Kloet and L. van Lierre (1990b). First attempt to apply whole-lake food-web manipulation on a large scale in the Netherlands. *Hydrobiologia* **200/201**, 291–301.

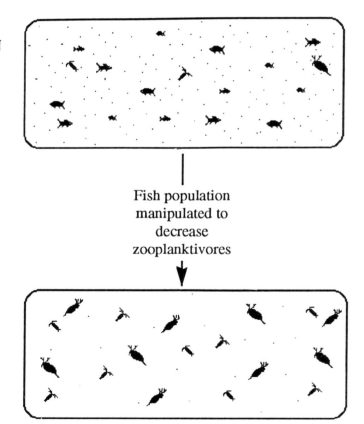

PRE-BIOMANIPULATION
1 many zooplanktivorous fish
2 many phytoplankton
3 few zooplankton

Fish population manipulated to decrease zooplanktivores

POST-MANIPULATION
1 few zooplanktivorous fish
2 reduced phytoplankton
3 more or larger zooplankton

Figure 1. Changes in zooplankton and phytoplankton populations in a waterbody following biomanipulation involving the elimination or reduction of zooplanktivorous fish.

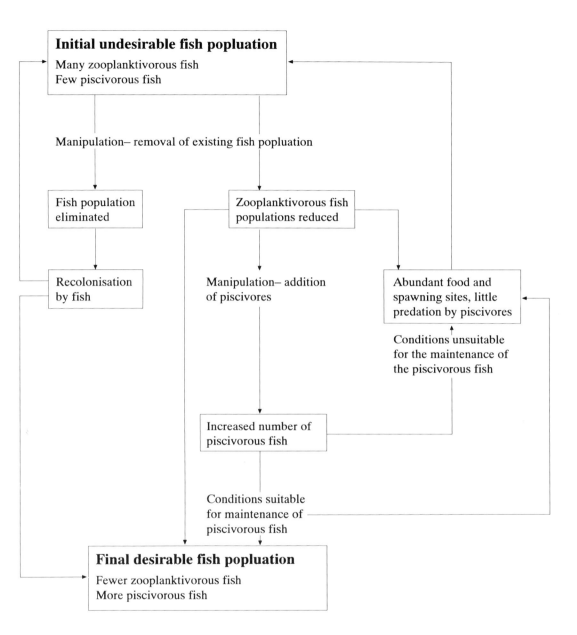

Figure 2. Likely outcomes following the manipulation of existing fish populations (for explanation see text).

ENHANCEMENT OF ESTUARINE HABITATS IN ASSOCIATION WITH DEVELOPMENT

R.M. Morton
WBM Oceanics Australia
PO Box 203
Spring Hill QLD 4004

Introduction

Urban development is proceeding rapidly in sheltered estuarine and coastal areas of sub-tropical Australia. This trend is unlikely to slow. Fringing the estuaries there are large areas of agricultural lands that are of low ecological value, and increasingly these areas are becoming used for residential development. Most development scenarios involve the creation of canals or lakes because these lands are low-lying and require filling to raise them above flood/storm surge levels. This style of development is also related to the recent greater demand for waterfront tourism areas and recreational amenities such as marinas. The increased population residing in estuarine regions results in waterways being used to a greater extent for stormwater disposal, floodwater conveyance and recreational activities (eg. angling or boating) often leading to a need for dredging operations.

There is a need to ensure that, if such developments and associated requirements are considered to be of community benefit, they are undertaken in a manner that not only minimises impacts on existing habitats but provides the opportunity to create new or additional habitat. Conversely, if the developments are judged to result in adverse ecological impacts then there should be a requirement to compensate by creating habitat (ie. the "no net loss of estuarine productivity principle").

The problem is, if we are to require compensatory works or enhancement works, how do we design them and how do we assess the contribution of any artificial waterways created by development to the ecology of the estuary? Unfortunately, there are very few Australian studies on creating fish habitats or examining fish fauna in artificial habitats such as marinas, canals, or similar developments (Saenger and McIvor 1974; Morton 1989; Morton in press; Hair and Bell in press). We know that it should be possible to create suitable conditions for estuarine fauna and to design developments appropriately, but opportunities are only recently becoming available as environmental considerations achieve greater prominence.

Larger scale developments or operations offer an opportunity to enhance or create aquatic habitat. The amount of money involved in such projects is such that even a small percentage of the total capital investment can be sufficient to fund the creation of large areas of habitat either associated with the development or in degraded areas external to the development.

Habitat enhancement/creation schemes that can result from large scale development include (1) those which represent an integral part of the development, and (2) works which are undertaken as compensation for disturbance of ecologically important habitats.

The following discussion describes (i) a study (Tweed Heads Notional Dredge Plan and River Management Study) designed to minimise adverse impacts to aquatic fauna resulting from required extensive dredging works and (ii) opportunities to provide additional habitat in conjunction with private waterfront developments.

Specific case studies

(i) Tweed River Plan of Management

The lower reaches of the Tweed River on the Queensland/NSW border have been shoaling over the past two decades, primarily as a result of infilling with oceanic sand, and this trend is expected to continue. There is a need to dredge some areas of the estuary to overcome significant navigation, flooding and potential water quality problems. For example, the entrance sand bar is very shallow, substantially affecting navigation, and could force the 25 trawlers based at Tweed Heads to abandon the port. Periodic flooding often causes significant damage to developed areas fringing the Tweed River. The dredging works would need to be extensive, and could result in the extraction of several million m^3 of sand. If the sand could be sold, from $7 million to $24 million could be raised in royalties depending upon the extent of dredging.

WBM Oceanics was commissioned to collaborate with the New South Wales Public Works Department to devise a Notional Dredge Plan to undertake necessary dredging (which forms part of the River Plan of Management). A primary aim of the Notional Dredge Plan is to ensure that dredging works and associated environmental compensatory/enhancement measures result in a net improvement in the overall ecological wellbeing of the Tweed River estuary.

The major ecological concern in regard to required dredging relates to hydraulic changes resulting from dredging works which will include an increase in the tidal range of the lower estuary. Depending on the degree of change, lower low tides could increase the aerial exposure of seagrass beds and perhaps result in seagrass losses. Alternatively, the higher high tides may increase mangrove extent.

The Notional Dredge Plan was designed to ensure that the most productive areas of the lower Tweed estuary are not directly disturbed. Dredging was restricted from occurring in river bank areas shallower than 2.0 m Australian Height Datum (AHD), as vegetation, bird, benthic and fish studies indicate that shallow shoreline areas (particularly bays/inlets) have a high ecological value. No seagrass beds would be dredged and buffer distances were defined around mangrove, seagrass, commercial net hauling grounds and oyster leases areas. Emphasis has been placed on avoiding potential future adverse erosion/scouring processes within the estuary which could require remedial dredging operations. Dredging schedules were designed to minimise and "stagger" alterations to the tidal regime of the estuary, whilst recognising the beneficial aspects of dredging.

The Notional Dredge Plan included designs to enhance dredged and existing habitats as well as options to create new habitat to benefit the ecology and fisheries of the Tweed River. Funds for such works would result mostly from royalties arising from the sale of extracted sands.

Dredging depths, as defined by the Plan, would not be uniform but would be contoured to provide habitat diversity. In some reaches of the river, several deep holes (to -9 m AHD) were included (Figure 1) to provide suitable habitat for large pelagic fish and thus enhance angling in these locations. Dredging of the Tweed River entrance is likely to increase the occurrence of pelagic and reef dwelling fishes in the estuary because access for these oceanic species will be easier. The Plan therefore has proposed to place artificial reefs in some areas (Figure 2) to provide habitat suitable to "attract" fish and provide additional angling opportunities for recreational fishermen. Computer modelling studies were undertaken to ensure that dredged areas would maintain high water-quality conditions.

Dredging strategies would include retention, and in some cases creation, of shallow sloping banks to provide areas for net hauling by commercial fishermen or for seagrass establishment. Anticipated water quality conditions (eg. current velocities, substrate type) in areas created using dredge spoil were computer-modelled to ensure that created conditions would be suitable for seagrass colonisation or transplantation.

Letitia Spit, part of Fingal Peninsula which forms the eastern bank of the lower Tweed River (Figure 3), was identified as being particularly suitable for habitat creation because most of the area concerned is Crown Land isolated from densely populated urban areas by the Tweed River. The Spit has important recreational values but presently mostly supports degraded vegetation as the area was previously used for sand mining operations. Several tidally connected lagoons (eg. Kerosene Inlet, Figure 3), remnants of previous mining operations, were identified during baseline ecological studies as supporting large areas of seagrass and diverse fish/prawn populations.

The Tweed River Plan of Management therefore provides designs (Figure 3) to protect these lagoons and create additional similar lagoons to enhance the value of the lower estuary to aquatic fauna including species of fisheries importance. Detailed ecological, water quality and hydraulic studies of existing lagoons were completed to assist in the design of proposed additional lagoons. Particular emphasis would be placed on providing conditions suitable for seagrass establishment to compensate for anticipated seagrass losses occurring as a result of dredging-induced changes in the tidal regime (see above). These works, and associated habitat enhancement projects (eg. creation of bird habitat) would be funded from royalties resulting from extraction of sands occurring within the river and from Letitia Spit.

Several ecological baseline studies (benthos, mangrove and seagrass) have been completed and future monitoring studies have been designed to enable firstly, the rapid identification of any substantial impacts that may result from proposed dredging, thus enabling modification of techniques and quantification of substantial impacts where they are unavoidable (eg. alterations to the hydraulic regime of the river) and secondly, to verify the need for and scale of compensatory works (eg. habitat creation/enhancement). Monitoring studies would also include created or enhanced habitats to validate the success of such works.

The Lower Tweed River Plan of Management includes a variety of habitat creation/enhancement schemes of a magnitude that could only be achieved as a result of a large scale operation. These options could be funded as a result of dredging royalties with no cost to the community which would also benefit (eg. in terms of flood relief, enhanced recreational opportunities and improved navigation) from proposed dredging works.

(ii) Artificial habitats associated with residential development

The view that all canal developments provide "poor quality habitat" and are detrimental to an estuary is generally based upon the performance of earlier designs. However, the level of planning and design expertise is far greater than in the past and there is no reason why a canal or tidally flushed lake style development cannot be designed to provide habitat suitable for supporting large fish populations of considerable ecological value.

There are several existing canal developments that have "accidentally" provided habitat that would otherwise not be available within a natural estuary. The author has seen large areas of seagrass (*Zostera capricorni*) in residential canal developments at Forster, Port Macquarie, Yamba and Tweed Heads (NSW), and numerous other canal developments also presumably contain seagrass. Some canals provide good angling opportunities (Morton in press) and large fish often shelter in deeper regions of canals. In many cases, commercial fishermen have netted in canal developments (much to the concern of residents fringing the canals) because of the large populations of fish present. Floating structures (such as pontoons) associated with marinas provide a unique non-tidal substrate subject to continual high light intensity and are often utilised by large numbers of juvenile fish (Hair and Bell in press), including species of fisheries value. Similarly, there are many examples of mangroves colonising man-made habitats such as stormwater drains and revetment walls. In some instances mangroves have colonised man-made structures in regions where they would not normally occur and thus enhanced the region's physical diversity and food resources for aquatic fauna.

These examples of "accidental" habitats demonstrate that habitat of value to fish, such as mangroves or seagrass, can be incorporated into waterway-orientated residential development. Mangroves could provide an attractive landscaping/screening feature of many developments. However, mangroves are protected and a permit is required to cut or lop mangrove trees. Existing legislation could be modified to protect already established views by allowing the pruning/lopping of mangroves (as would be required in such situations) which have been artificially established.

Future developments could include design attributes that specifically aim to provide habitat for fish, either within the development itself, or in nearby areas specifically dedicated for this purpose. These attributes could form a requirement of approving authorities or be included by the developer to improve the public image of the development. In some instances, man-made tidally-flushed lakes or canals constructed in non-tidal areas may be beneficial to an estuary in that they provide additional aquatic habitat. Similarly, a requirement for the creation of artificial fish habitat as compensation for the loss of small areas of natural habitat (eg. as a result of entrance channel requirements) may allow a development to be constructed without adverse ecological effects. For example, previous urban development of the catchment in many estuaries has not incorporated measures to reduce/prevent elevated sediment loads in stormwater runoff. This has lead to the siltation of many shallow creeks and bays to the detriment of the estuary. Properly planned dredging can be undertaken to restore waterway depths and therefore enhance the habitat value of many degraded areas.

Overview

Development of many areas fringing estuaries is inevitable. Large scale development offers an opportunity to plan appropriately and adopt compensatory measures where necessary. In large developments, the scale of capital expenditure is such that habitat enhancement, creation and preservation schemes (eg. dedication of parks/reserves, creation of artificial wetlands) are a realistic proposition. The piecemeal type of smaller scale development offers few such opportunities and provides little hope for conservation of estuarine resources unless long term Estuarine Management Plans are enacted and reinforced by regional town planning laws. Basic information is available to define conditions required to provide habitat of value to fish although there is a need for further studies to identify the best design attributes of artificial habitat and how such habitat can be incorporated into developments.

Acknowledgements

I thank the New South Wales Public Works Department for permission to use information from the Tweed River Plan of Management. Mr Brian Dooley (Public Works Department) provided important guidance and assistance in formulation of the Notional Dredge Plan.

References

Hair, C.A. and J.D. Bell (in press). Effects of enhancing pontoons on abundance of fish: Initial experiments in estuaries. *Bulletin of Marine Science.*

Morton, R.M. (1989). Hydrology and fish fauna of canal developments in an intensively modified Australian estuary. *Estuarine and Coastal Shelf Science* **28**, 43-58.

Morton, R.M. (in press). Fish assemblages in residential canal developments near the mouth of a subtropical Queensland estuary. *Australian Journal of Marine and Freshwater Research.*

Saenger, P. and C.C. McIvor (1974). Water quality and fish populations in a mangrove estuary modified by residential canal developments. *Proceedings of the International Symposium on Biology and Management of Mangroves* **No. 2**, 753-765.

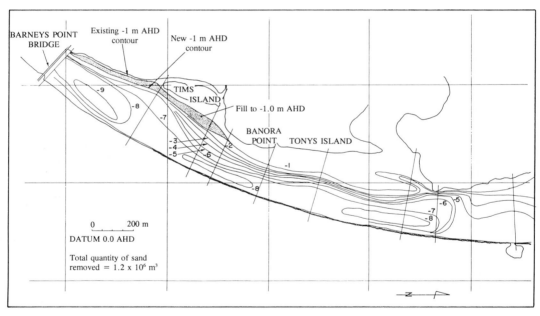

Figure 1. Example of part of Tweed River Notional Dredge Plan indicating areas proposed for filling (to create conditions suitable for seagrass establishment) and areas where deep holes will be excavated (to enhance fish abundance).

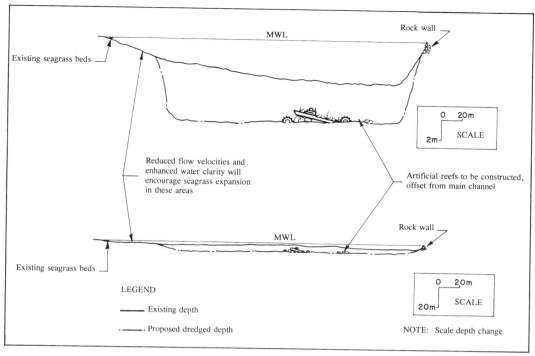

Figure 2. Typical schematic cross-section of main Tweed River channel following dredging (NB differing depth scales).

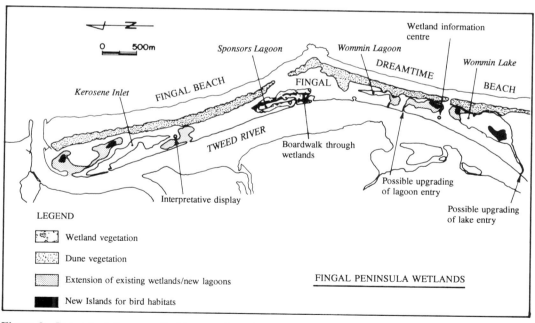

Figure 3. Conceptual diagram of habitat creation scheme for Letitia Spit (Fingal Peninsula).

DISCUSSION OF SESSION 3

Recorded by G.A. Thorncraft
Fisheries Research Institute
P.O. Box 21
Cronulla NSW 2230

Each presentation from the panel was followed by a brief period for questions, after which the floor was opened for general discussion.

Following *Ian Poiner's* presentation, Campbell Davies asked how quickly the seagrass beds recovered after cyclone damage. Ian Poiner replied that recovery of the seagrass beds to their former area may occur in ten years, but recovery of the fisheries was another matter.

Duncan Leadbitter suggested that management of seagrass beds for prawn fisheries was a classic case of single species management, and asked how a more holistic ecosystem approach to management may differ? An ecosystem-based approach was definitely needed according to Ian Poiner, with the qualifier that a fisheries' perspective to management should be maintained to ensure the fishing industry is not disadvantaged under an ecosystem approach to management.

Stan Moberly suggested to *Stephen Swales* that rehabilitation was a step in the right direction, but that not messing it up in the first place was preferable. Stephen Swales agreed that proactive measures are desirable, but depending on the situation one has to be proactive or reactive.

Murray MacDonald was concerned that habitat restoration was being seen as an approach that legitimises continuing degradation of habitat and asked whether we ought to be pushing this technique? Stephen Swales agreed that restoration was a last resort and that we should promote proactive measures wherever possible, but given the reality of the situation we have to be prepared to have tools available to mitigate impacts of development on fish habitats.

George Paras commented to *Mick Bales* that biomanipulation models to establish desirable fish communities were overly simplistic because of the complex interactions between prey groups at lower trophic levels. Mick Bales replied that this was one of the reasons why it is difficult to predict the outcome of any biomanipulation exercise with accuracy.

Karen Edyvane asked Mick Bales if he had thought about modelling physical processes along with the food web information. He responded that he thought other people had tried it, but he hadn't modelled physical processes himself.

Hugh Cross likened the example *Rick Morton* presented to a management experiment and suggested that it is often advantageous to manipulate a system and see how it responds. While this example covered the responses of the Tweed estuary, he questioned whether there had been a change in sand distribution either upstream or farther up the coast? Rick Morton answered that there were four components to the study which addressed water quality; estuarine flora-fauna; terrestrial flora-fauna; and social economics. The social economics study looked at the demand for sand over the long term and took into account such things as stock piling sites and the impact of extra traffic resulting from sand extraction.

John Glaister asked whether the study addressed more northerly habitats which had been degraded by the prevention of sand migrating up the coast? Rick Morton responded that the entire problem lay in the design of the breakwaters, and that sand would continue coming into the Tweed estuary until a bypass system was constructed to allow currents to transport sand farther north. Until that time, the need for 'remedial' dredging of the estuary will continue. The current dredging plan assumes that a sand bypass system will be put in and the channels are all dredged to an equilibrium situation so they should not accrete or erode in any areas.

The *general discussion* started with Jim Puckridge commenting that temporal variability was an important issue particular to Australian rivers and was a major factor in structuring fish communities. He asked Stephen Swales about approaches to restoring temporal variability to the hydrological regime. Stephen Swales replied that natural variability was a component of the environmental flow program he was working on to design suitable flow regimes for the regulated rivers in NSW.

Peter Jackson asked Stephen Swales if he saw value in the concept of having a volume of the impoundment allocated to environmental purposes, rather than a minimum flow. The environmental agency responsible for the impoundment would be able to manipulate its share of the water independently of the water management agency. Stephen Swales responded that NSW Water Resources (DWR) was examining environmental contingency allowances within its storages. Hugh Cross added that DWR has a three component environmental flow policy including an operations component where water is released from dams for irrigation, an environmental contingency component which can act as a flush or piggyback release on tributary flows, and an unregulated flow component. This system has to be flexible because dams are small and have an insignificant effect on flood flows for instance, whereas others hardly ever fill because the dam is too big for the catchment.

Murray MacDonald repeated his earlier concern about using habitat enhancement and restoration as means of legitimising continuing disturbances of habitats. He asked Rick Morton if he believed that proposals to re-create habitats were actually going to come anywhere near compensating for the loss in productivity since the beginning of significant European impact, and if not, how can we claim that enhancement and restoration is actually looking after habitats *in toto* and not just a piecemeal effort. Rick Morton apologised for not having data preceding European settlement, but reiterated that the Tweed River needs to be dredged to prevent the estuary clogging with sand. While dredging may not totally compensate for the sand intrusion, it was certainly a step in the right direction.

Barbara Richardson asked what follow up work would be undertaken in the Tweed River program, and other similar research programs. In this case, Rick Morton said it was up to the NSW Public Works Department to commission follow up work, however he suggested that if a developer is required to put in compensatory works, then there should also be an accompanying requirement to monitor them. As an example, a number of American states require developers to put in compensatory salt marshes, with successful compensation being determined by a minimum percentage of sprig success and regeneration.

Bob O'Boyle agreed that when developing a model, follow up work is essential to find out if that model really does reflect reality. He commented that the dredging of Halifax Harbour uncovered a Pandora's box of heavy metals and other leaching agents in the sediments and asked Rick Morton whether similar aspects were addressed in the Tweed estuary. Rick Morton indicated that there was very little silt associated with the clean oceanic sand that was entering the estuary, so turbidity plumes and related problems were unlikely to occur. However silt and pesticide analyses were undertaken as a precaution.

Derek Staples asked Ian Poiner if he remembered 15 years ago wearing T-shirts reading 'Muddies Need Mangroves', and then questioning whether mud crabs really needed mangroves at all? He then asked if mud crabs still needed mangroves and if so why? Ian Poiner strongly denied ever wearing those T-shirts for fear of offending Dr Hill, but believed that mud crabs don't really need mangrove systems. While agreeing with ecosystem approaches to coastal management, he thought a much more focused attitude should be taken to identify the key factors which sustain the fishing industry. He believed the issue ultimately comes down to questions of allocation between developers and fishermen, and didn't really think the fishing industry was in a strong position at this time to win many allocation battles.

Peter Young expressed concern about the religious zealotry that seemed to go around every time someone said 'habitat'. He also put forward the idea that the ecosystem must be modified to maximise the yield of fish.

Ian Poiner commented that fishery scientists still do not really understand the couplings between fish populations and shallow water coastal habitats. He said that it was illogical to manipulate a poorly-understood system and somehow 'twiddle' it to maximise a few populations that have value to the fishery, because that is a very risky strategy. Rob Lewis argued that fisheries agencies needed to integrate all the issues and considerations involved in traditional fisheries objectives to provide what society expects today.

Campbell Davies commented that recent work at James Cook University shows there is a tidal coupling between juvenile reef fish and seagrass and mangrove areas. Ian Poiner reflected that there is no doubt that mangroves and seagrass are important to fisheries, but he would be very surprised to find that all seagrasses are important to those juvenile fish. Rick Morton further remarked that not every mangrove is important, as some mangroves growing in areas that are barely touched by the tides couldn't possibly have the same ecological value as mangroves that have the tide coming in twice a day with fish feeding amongst them. Jenny Burchmore expressed her concern that this was a really dangerous area to get into from a management perspective, because unless each mangrove stand and seagrass bed in each State is investigated to determine its ecological function, these areas need blanket protection to protect the vitally important areas that we do not yet understand. To indicate that some of these areas were less important than others was a very dangerous game to play.

Ian Poiner continued the debate by saying he would like to see those activities impacting seagrass systems, such as the northern prawn fishery, stopped, but couldn't see it happening in the foreseeable future because of the question of resource allocation. Rick Morton agreed and, using a mining project as an example, said that the federal government had decided that the benefits of mining to the Australian community outweighed the loss of an area of wetland. The pragmatic question was then to minimise the impact of mining on wetlands.

Jeremy Prince questioned what people were actually fighting to maintain. He explained that in British Columbia, an abalone fishery was recently closed down to protect declining stocks as a conservation issue. The interesting thing was that the abalone fishery was only a recent development since sea otters, which kept abalone populations in check, had been virtually hunted to extinction in the area before the turn of the century. Thus the modern abalone fishery was exploiting a resource which came into being because sea otters were no longer there. Now the abalone fishery has been closed down on conservation grounds, and the sea otters are coming back, abalone stocks appear to be reverting to historical background levels. Jeremy Prince's point was that whatever we try to sustain, we need to be very clear on whether we want to preserve a pristine environment, or maintain a productive environment for human ends?

SESSION 4

ORGANISMS AND ENVIRONMENTAL RELATIONSHIPS II — KEY VARIABLES AND BROAD BASED ISSUES

Session Chairperson: B.E. Pierce
Session Panellists: J.D. Koehn
P.C. Young
N.R. Loneragan
Rapporteur: M.I. Kangas
Disccusion of Day 1: D.C. Smith

CHAIRPERSON'S INTRODUCTION: KEY FACTORS LINKING FISH AND THEIR HABITAT

B.E. Pierce

Inland Waters Section
South Australian Research & Development Institute
GPO Box 1625
Adelaide SA 5001

If we define the habitat of a fish simply as that subset of the environment where it makes its living, it is intuitively obvious that fish populations are inextricably linked to the habitat they call "home." People change fish habitat but still want a healthy resource. Therefore, identifying and measuring the key factors linking fish and habitat becomes the cornerstone of management for sustainability (eg, through research direction, monitoring, regulation and public education).

This conceptually simple linkage process is complicated by (amongst other issues):

- the multivariate nature of both the dependent and independent variables (ie, community, fish population, habitat);

- uncontrollable variability in both the habitat and resource;

- multiple, often unidentified human influences (treatments);

- time-treatment interactions;

- the reactive (as opposed to proactive) nature of monitoring-based assessment;

- a lack of clear societal vision/definition of the desired goal (eg, do we want sustainability of stocks at 50% of natural levels, 25% with some extinctions, 95% with no extinctions?) for most fish stocks/communities;

- the fact that we are already dealing with a modified system (initiated experiment) in most cases; and

- limited funding and resources.

While limited research has been undertaken on specific factors, the relative importance of each has not generally been evaluated. Ryder and Kerr (1989) have attempted a first approximation by erecting four over-riding determinants at the environmental level:

- dissolved oxygen
- water temperature
- subsurface light
- dissolved nutrients

The usefulness of these in non-salmonid communities has yet to be evaluated at a community or ecosystem level to predict the impacts of anthropogenic habitat change.

Postscript to the workshop

We might have expected a conceptually simple series of outcomes from this session: tentative key factors by biotope together with a list of useful methods. In reality, discussions were disparate with no clear "take home message." Only well after the workshop have summary threads come together in my mind. I offer the following personal observations:

- While there appeared to be consensus going into the workshop that habitat was critical to fish, it is clear that in no system (freshwater, marine, or estuarine) have we even begun to understand or even closely examine the linkages which must support the associated resources. Put another way, there is a lack of evidence to strongly reject the null hypothesis of no linkage between fish and habitat.

- While we are good at listing general factors which are likely involved in such presumed linkage, quantification of most is lacking nationally and globally. To me it was surprising that relatively few biologists held strong views as to what the key factors were in their chosen ecosystem. Apparently this research issue is far from the focus of their thinking.

- Human habitat impacts appear, based on current knowledge, to have the greatest treatment effect in limited freshwater systems with progressively less influence through estuarine systems to coastal and finally pelagic oceanic ecosystems.

- Little consensus exists as to "best" methods for this research arena. Multivariate analysis appears the most rigorous and efficient, particularly where treatments can be manipulated in real systems (eg, Walters 1986; Spellerberg 1992). However, much of the specific information discussed during and after the session fell into the category of "observations" resulting in mental "correlations."

Awakening in a cold sweat some days after the workshop, I had a vision of old Chris Columbus pointing an accusing finger at me and calling me a "flat earth biologist" who still insisted that the world rotates around fish, rather than waking up to the reality that fish rotate around their habitat. Perhaps the current fisheries legislation throughout Australia which focuses primarily on harvest control rather than habitat control is symptomatic of a similar paradigm. Decline trends in many world fisheries and particularly inland systems seem to indicate that the current treatments aren't particularly successful. However, as with other paradigm shifts, habitat/fish linkage may not become evident until we refocus on habitat management as the cornerstone of resource sustainability, for example, as the European Economic Community is initiating (Mader 1991).

References

Mader, H.J. (1991). The isolation of animal and plant populations: Aspects for a European Nature Conservation Strategy, 265-275. *In* A. Seitz and V. Loeschcke [Eds.] *Species conservation: A population-biological approach.* Birkhauser Verlag; Basel, 281pp.

Ryder, R.A. and S.R. Kerr (1989). Environmental priorities: placing habitat in hierarchic perspective, 2-12. *In* C.D. Levings, L.B. Holtby and M.A. Henderson [Eds.] Proceedings of the National Workshop on effects of habitat alteration on salmonid stocks. *Can. Spec. Publ. Fish. Aquat. Sci.* **105**.

Spellerberg, I.F. (1992). *Evaluation and assessment for conservation. Ecological guidelines for determining priorities for nature conservation.* Chapman and Hall; London, UK, 260pp.

Walters, C. (1986). *Adaptive management of renewable resources.* MacMillan Publishing Company; New York, New York, USA, 374pp.

FRESHWATER FISH HABITATS: KEY FACTORS AND METHODS TO DETERMINE THEM

J. D. Koehn

Freshwater Ecology Section
Department of Conservation and Natural Resources
123 Brown St.
Heidelberg VIC 3084

Introduction

As fish habitat generally remains hidden underwater, is somewhat foreign to humans as terrestrial beings, and can be difficult to measure, it has often been neglected in the study and management of fish species and fisheries. As with terrestrial species however, freshwater fish species are dependant on the availability of habitat, and it is the alterations to, and destruction of, this habitat which are responsible for the decline in fish species and fisheries that have been witnessed. In general, freshwater fish appear more dependant on physical habitat than their marine counterparts, as freshwater ecosystems are more confined by their surroundings and are generally under more pressure from environmental changes, due to impacts on those surroundings. Although changes to stream environments and their catchments have been well documented in only a few areas (eg. Mitchell 1990), general environmental changes have long been recognised (eg. Merrick and Schmida 1984; Lake and Marchant 1990) and the causes of these changes and threats to fishes from them extensively discussed (Cadwallader 1978; MDBC 1988; Koehn and O'Connor 1990a; Jackson *et al.* in press). It is the changes to the key habitat factors which are responsible for the general decline in fish species.

Key habitat factors

Key habitat factors essentially comprise of the water and its condition, the surrounding land which helps determine that condition, conditions that the water creates within the stream, and instream objects.

1. Water

Water is the principle component of fish habitat and whilst much public attention focuses on quality the primary concern must be the amount of water that is present.

Quantity

The amount of water determines the area of the streambed which is covered (usually measured by wetted area or wetted perimeter) and hence the amount of aquatic habitat available. The amount of habitat area does not necessarily increase proportionally with flow however, and a combination of particular habitat variables which may be deemed as important for a species may even decrease as higher flows increase water velocities (Tunbridge and Glenane 1988). Examples of such changes are given in Figures 1 and 2.

The amount of water can also determine whether particular habitats actually receive water or not, eg. overbank flooding into billabongs.

Flow regime

The timing of the amount of water in the stream determines the flow regime. This incorporates small, short-term variations, large variations (eg. floods), seasonal variations, the timing of such variations, and the rate of changes in flow.

Water velocity

In lotic ecosystems, water velocities are an important habitat factor. The habits and swimming capabilities of the species help determine the preferred water velocity. Variations in velocity, and shelters from fast velocities provided by variations in the substrate, substrate particles or other instream objects, are important in providing suitable habitat and refuges from other species. Water velocity generally decreases with depth.

Depth

Water depth can determine the amount of a particular habitat variable (eg. wood debris) which is available in the water column. Vertical space in the water column is likely to be of greater need to mid-water schooling species than benthic species. Water depth is an important factor in avoidance of terrestrial predators, and in conjunction with water velocity it determines the stream habitat type eg. pool or riffle. Depth and velocity gradients provide the major component for fish microhabitat use in many stream fish assemblages (eg. Grossman et al. 1987a, b; Angermier 1987). Depth can provide relatively stable, sheltered areas whereas shallow areas are particularly sensitive to reductions in water levels. With increasing depth, light penetration decreases and hence visibility is reduced, providing protection from predators.

Quality

Acceptable water quality is an essential prerequisite for fish habitation and each species has different tolerance levels to different water quality parameters (Koehn and O'Connor 1990b). Some of the major water quality criteria are:

a) Suspended sediment: Can kill fish at high levels, can smother eggs, cause stress and affect feeding (Newcombe and MacDonald 1991). It can also cause invertebrate drift (Doeg and Milledge 1991) and ultimately smother substrates (Berkman and Rabeni 1987).

b) Temperature: Often a forgotten water quality criterion, the water temperature controls the metabolic rate of the ecosystem and the fish in it. Each species has upper and lower temperature tolerance levels as well as specific requirements for activities such as spawning (Koehn and O'Connor 1990b).

c) Dissolved oxygen: Suitable levels are essential for respiration. Stream levels can be altered by flow and temperature, and by instream structures such as riffles which assist in aeration.

d) Salinity: Whilst high levels can preclude freshwater fish from some habitat areas (Anderson and Morison 1989), appropriate salinity levels can also be important for those species which have an estuarine or marine phase to their lifecycle.

e) Nutrients : Whilst nutrients are an important factor in stream production, excessive nutrient loads are a major contributing factor in eutrophication.

f) Toxins: Can cause death, stress or increased abnormalities.

There are also a large number of other chemical parameters which may influence the needs of certain fish species and to which species will have varying tolerance levels or requirements (eg. pH, calcium).

All of these water quality parameters can be affected by the amount of flow. Reduced flows can exacerbate toxic effects through reduced dilution.

Rivers and streams flow in one direction and so point source changes to the water quality have the potential to affect the system for large distances downstream.

Habitat diversity in the form of changes in depth, velocity, temperatures and other chemical parameters are all important, providing the needs of different species and lifestages and maintaining the heterogeneity of components in the ecosystem.

2. Surrounding habitat

Catchment
Catchment management has a major influence on the water quality of the stream and also influences the quantity of water available. Runoff is affected by activities such as urbanisation, drainage, deforestation and reforestation. Any activity in the catchment has the potential to affect the stream water.

Banks
The bank is important as it forms the perimeter of the stream, is the boundary between the water and the land, forms the stream channel and keeps the water within the stream. This helps determine water depth and velocities, available habitat and habitat variation.

Riparian zone
This vegetation zone is essential to the well-being of aquatic ecosystems and has continual interactions with the stream. It acts as a buffer between the surrounding activities and the stream, filters runoff, provides shade and inputs of organic material. In upland streams these inputs provide the major energy source in the form of leaves, bark etc, and much of the instream habitat (snags, logs, branches) originates from this zone. The root systems of this vegetation also play a major role in erosion prevention.

Shade
This can be particularly important to those fish species which avoid sunlight or wish to escape predators and is also a factor in the prevention of algal blooms.

3. Instream habitat
Instream objects are important in providing structure to the underwater habitat. They provide shelter from water velocity and sunlight, orientation points around which a territory or habitat area can be based, spawning sites, and areas in which to hide from predators or hide in wait for prey. Such objects fall into three main categories:

a) Substrate: Substrate particles provide important refuge areas for small fish and juveniles whilst substrate undulations provide particular habitat areas through variation in depth.

b) Wood debris: Fallen trees, branches, logs, (all usually referred to as snags), and associated organic debris, form a major component of instream habitat, especially in lowland rivers. In addition to providing the advantages mentioned for instream objects they cause variations in flow and depth, provide habitat areas and spawning sites for species such as freshwater blackfish *Gadopsis marmoratus* (Koehn 1986; Koehn et al. in press; Jackson 1978), and attachment sites for invertebrates (O'Connor 1991).

c) Aquatic plants: Provide habitat and spawning sites for fish and attachment sites for invertebrates, bed and bank stability and shade.

As each species has its own particular habitat needs, and these may change throughout its lifecycle, a diversity of habitat is essential to provide these needs. This diversity is provided in part, by a diversity of other attributes such as flows.

Access to habitats
For available habitat to be of use it is essential that fish species have access to it. Harris (1984) calculated that about 50% of available habitats in south-eastern New South Wales and far eastern Victoria was not available to migratory species. As about 70% of the fish species in these coastal drainages need to migrate between freshwater and the estuaries or the sea at some stage of their lifecycles (Koehn and O'Connor 1990a), barriers to fish passage mean that a large

proportion of the available habitat may not be available to the majority of fish species in this region. Such barriers can be formed by major structures such as dams, weirs, retarding basins, etc, or even by less obvious structures such as poorly designed culverts and road crossings. In addition to species which are known to make large scale migrations, all species need to be able to move freely within the stream in order to recolonise and find suitable habitat areas, mates etc, and there are many species for which movement needs are completely unknown. Serious attention needs to be given to redressing the problems of barriers to fish passage as a habitat issue.

Other influencing factors

There are several other biotic factors such as introduced species, diseases and food supply which can have a major influence on fish populations, and although they can often act independently of the habitat available, their effects can be influenced by it. For example, introduced species with wide habitat tolerances such as carp *Cyprinus carpio* may be advantaged over native species which have particular requirements where those requirements have been degraded. Additional stress caused by suboptimal habitats may assist diseases, and lack of food supply could well reflect general stream habitat degradation. Such factors are well canvassed in other publications (Pollard 1989; Morison 1989; Hynes 1970).

Methodologies to determine habitat linkages

There is still an urgent need for appropriate basic biological information on freshwater fish. A collation of all biological information on Victorian native freshwater fish (Koehn and O'Connor 1990b), shows that there are many gaps in our knowledge base. In order to manage any species or ecosystem properly, knowledge of the ecology of the organisms is essential.

With a limited number of researchers and funds, it is imperative that the best use be made of the resources available to provide the best information for managers. This means that before a study is undertaken the most effective method of research must be determined. The first question that must be asked is: What is the required outcome? What is the information needed for? What is the question that we are going to try to answer? Then, subsequent questions such as: What degree of resolution do we need to answer the question? Is a yes\no answer enough or do we need further details?; what degree of scientific rigour do we need to achieve?; what are the parameters that need to be measured?; what parameters can be managed?

These will assist in determining the method best suited to answering the question at hand.

Methods

Multiple parameter studies

- Wide-scale studies measuring habitat characteristics at several rivers within a region and correlating them with fish distributions and numbers at those sites (eg. Davies 1989).

- Localised studies using instream sections where habitat characteristics are measured and correlated against fish numbers (eg. Koehn 1986; Koehn *et al.* in press).

- Fish position studies which use the instream locations of individual fish determined by site of capture, observation, radiotracking, etc. (see Grossman and Freeman 1987; Tyus *et al.* 1984). This type of study is often used to determine frequency of use curves and habitat suitability models (eg. Terrell 1984).

All of these methods measure a range of habitat parameters and correlate fish numbers against these parameters using regression analysis, principle components analysis, etc. (Zar 1984; Digby and Kempton 1987).

Single parameter studies
These determine the importance of only one parameter and may often be undertaken as experimental studies in the field or laboratory, eg. LC_{50}, tolerance tests (Bacher and O'Brien 1988).

Single aspect studies
Study one particular aspect of the fish and determine the requirements for that aspect eg. the spawning of freshwater blackfish *G. marmoratus* (Jackson 1978).

Observations
Field or laboratory observations may provide answers to simple questions such as the nature of the species, and may give a good indication as to the direction that the study should take, and the methods to be used (Eldon 1969).

Distributional data
May give general information on the areas that the species inhabits and may be used in conjunction with other data to indicate areas of usage (eg. altitude ranges, barriers overcome).

Monitoring
Widespread monitoring undertaken for other purposes may not ideally be suited to obtaining direct habitat linkages. Such monitoring can however provide baseline data for the future and may indicate trends of populations. Monitoring of this type needs to be planned carefully to obtain maximum benefit from expenditure.

Monitoring the outcome of a particular event can provide direct information as to the effect on fish populations. There is an inherent possibility, however, that events requiring long-term monitoring (eg. timber harvesting) may not occur, funding may not be forthcoming (see Koehn *et al.* 1992), or that natural variation or chance events may cloud results.

Adaptive management
Manipulation of a 'natural' situation and monitoring of the effects can provide information on habitat requirements, for example, the placement of rocks in the Ovens River substantially increased the population of two-spined blackfish present (Koehn 1987).

In many cases several of these methods may have to be used together with physiological information to piece together the full picture of the requirements of the species. It should be remembered that this paper refers to management-oriented research that will provide appropriate answers to management in a time and cost effective manner. Academic theories and the study of freshwater processes remain important in the overall understanding of freshwater ecosystems and should always be considered when research is being designed and undertaken.

Constraints
When undertaking such research, apart from constraints imposed by resources, there are always constraints imposed by natural conditions which need to be taken into consideration. Some of these relevant to freshwater studies are: finding appropriate natural or relatively undisturbed habitats for study sites; obtaining sufficient fish numbers to be statistically viable; the effects of other (possibly introduced) fish species present; limitations imposed by the endangered or inaccessible nature of the species; barriers to fish passage downstream; fishing (legal and illegal); chance events (eg. a toxic spill, landslides); different life stages of the species or different habitats used at different times (eg. at night or during spawning); access to sites; the availability of efficient capture methods; river type; seasonal weather; or site conditions.

Concluding remarks
It is important that research conducted on the habitat requirements of freshwater fish species is directed toward the better management of those habitats. It is the alteration to their habitats that has caused the decline of most species and in most cases continues to be of threat (see Jackson *et al.* in press). These threats need to be addressed in terms of both research and management to prevent further declines in freshwater fish stocks. A positive step toward this is the

fish management plan for the Murray-Darling Basin (Lawrence 1991) which recognises and addresses many of these problems in that region, and similar plans for management and research are needed for other river basins.

Acknowledgements

Thanks to Bill O'Connor and the many other colleagues with whom endless discussions on this topic have taken place, and to Pat O'Leary who assisted with this manuscript.

References

Anderson, J.R. and A.K. Morison (1989). Environmental flow studies for the Wimmera River, Victoria. Summary Report. *Arthur Rylah Institute for Environmental Research Technical Report Series No. 78*. Department of Conservation, Forests and Lands, Victoria, 70pp.

Angermier, P. L. (1987). Spatiotemporal variation in habitat selection in small Illinois streams. In: Matthews, W. J. and Heins, D. C. (eds.). *Community and Evolutionary Ecology of North American Stream Fishes*. University of Okalahoma Press, Norman, 52–60.

Bacher, G. J. and T. A. O'Brien (1988). The sensitivity of Australian freshwater organisms to heavy metals. Environment Protection Agency Report, Melbourne, Victoria.

Berkmann, H. E. and C. F. Rabeni (1987). Effect of siltation on stream fish communities. *Environmental Biology of Fishes* **18**, 285–294.

Cadwallader, P.L. (1978). Some causes of the decline in range and abundance of native fish in the Murray-Darling River System. *Proceedings of the Royal Society of Victoria* **90**, 211–224.

Davies, P. E. (1989). Relationships between habitat characteristics and population abundance for Brown trout, *Salmo trutta* L., and Blackfish, *Gadopsis marmoratus* Rich., in Tasmanian streams. *Australian Journal of Marine and Freshwater Research* **40**, 341–59.

Digby, P. G. N. and R. A. Kempton (1987). *Multivariate Analysis of Ecological Communities*. Chapman and Hall, London, 206pp.

Doeg, T.J. and G. A. Milledge (1991). Effect of experimentally increasing concentrations of suspended sediment on macroinvertebrate drift. *Australian Journal of Marine and Freshwater Research* **42**, 519–26.

Eldon, G. A. (1969). Observations on growth and behaviour of Galaxiidae in aquariums. *Tuatara* 17, 34–36.

Grossman, G. D. and M. C. Freeman (1987). Microhabitat use in a stream fish assemblage. *Journal of Zoology (London)* **212**, 151–176.

Grossman, G. D., A., De Sota, M. C. Freeman and J. Lobon-Cervia (1987a). Microhabitat use in a Mediterranean riverine fish assemblage. Fishes of the lower Matarrana. *Oecologia* **73**, 490–500.

Grossman, G. D., A. De Sota, M. C. Freeman and J. Lobon-Cervia (1987b). Microhabitat use in a Mediterranean riverine fish assemblage. Fishes of the upper Matarrana. *Oecologia* **73**, 501–512.

Harris, J. H. (1984). Impoundment of coastal drainages of south-eastern Australia, and a review of its relevance to fish migrations. *Australian Zoologist* **21**, 235–250.

Hynes, H. B. N. (1970). *The Ecology of Running Waters*. Liverpool University Press, 555pp.

Jackson, P.D. (1978). Spawning and early development of the river blackfish, *Gadopsis marmoratus*, Richardson (Gadopsiformes:Gadopsidae), in the McKenzie River, Victoria. *Australian Journal of Marine and Freshwater Research* **29**, 293–298.

Jackson, P. D., J. D. Koehn and R. Wager (this meeting). Australia's Threatened Fishes. 1992 ASFB Listing.

Koehn, J.D. (1986). Approaches to determining flow and habitat requirements of freshwater native fish in Victoria. In: Campbell, I.C. (ed.) *Stream Protection. The Management of Rivers for Instream Uses*. (Chisholm Institute of Technology: Melbourne), 95–113.

Koehn, J.D. (1987). Artificial habitat increases abundance of two-spined blackfish *Gadopsis bispinosis* in Ovens River, Victoria. *Arthur Rylah Institute for Environmental Research Technical Report Series* No. 56. Department of Conservation, Forests and Lands, Victoria, 20pp.

Koehn, J. D. and W. G. O'Connor (1990a). Threats to Victorian freshwater fish. *Victorian Naturalist* **107**, 5–12.

Koehn, J.D. and W.G. O'Connor (1990b). *Biological Information for Management of Native Freshwater Fish in Victoria*. (Government printer: Melbourne), 165pp.

Koehn, J., T. Doeg and T. Raadik (1992). Aquatic Fauna. In: Squire, E. O. (ed.). First Interim Report for the Value Adding Utilisation System Trial 1989–1991. Dept. Conservation and Environment, Victoria.

Koehn, J. D., N. A. O'Connor and P. D. Jackson (in press). Seasonal and size-related variation in microhabitat use of a Southern Victorian stream fish assemblage (submitted to *Environmental Biology of Fishes*).

Lake, P. S. and R. Marchant (1990). Australian upland streams: ecological degradation and possible restoration. *Proceedings of the Ecological Society of Australia* **16**, 79–91.

Lawrence, B. W. (1991). *Fish Management Plan*. (Murray-Darling Basin Commission), 48pp.

Merrick, J. R. and G. E. Schmida (1984). *Australian Freshwater Fishes. Biology and Management*. Griffin Press Ltd., South Australia, 409pp.

Mitchell, P. (1990). *The Environmental Condition of Victorian Streams*. Dept. of Water Resources, Victoria, 102pp.

Morison, A.K. (1989). Management of introduced species in the Murray-Darling Basin—a discussion paper. In: *Proceedings of the Workshop on Native Fish Management*. Canberra 16–17 June 1988. Murray Darling Basin Commission, 149–161.

Murray-Darling Basin Commission (MDBC) (1988). *Proceedings of the Workshop on Native Fish Management. Canberra, 16–17 June 1988*. Murray-Darling Basin Commission, 1989.

Newcombe, C. P. and D. D. MacDonald (1991). Effects of suspended sediments on aquatic ecosystems. *North American Journal of Fisheries Management* **11**, 72–82.

O'Connor, N. A. (1991). The effects of habitat complexity on the macroinvertebrates colonising wood substrates in a lowland stream. *Oecologia* **85**, 504–512.

Pollard, D. A. (ed.) (1989). *Introduced and translocated Fishes and Their Ecological Effects*. Australian Society for Fish Biology Proceedings No. 8. Australian Government Publishing Service, Canberra, 181pp.

Terrell, J. W. (ed.) (1984). Proceedings of a workshop on fish habitat suitability index models. *U.S. Fish and Wildlife Service Biological Report* **85**(6), 392pp.

Tunbridge, B.R. and T.J. Glenane (1988). *A Study of Environmental Flows Necessary to Maintain Fish Populations in the Gellibrand River and Estuary*. Arthur Rylah Institute for Environmental Research. Department of Conservation, Forests and Lands, Victoria, 151pp.

Tyus, H. M. , B. D. Burdick and C. W. McAda (1984). Use of radiotelemetry for obtaining habitat preference data on Colorado squawfish. *North American Journal of Fisheries Management* **4**, 177–180.

Zar, J. H. (1984). *Biostatistical analysis*. Prentice-Hall, London, 718pp.

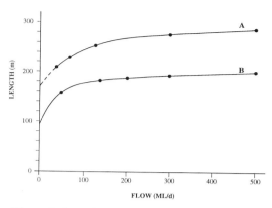

Figure 1. Length of wetted perimeter related to flow, measured across transects at two sites (A & B) on the Gellibrand river, Victoria. (from Tunbridge and Glenane 1988).

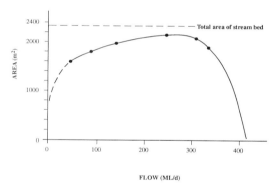

Figure 2. 'Rearing habitat' for freshwater blackfish *Gadopsis marmoratus* (defined as water with a depth >45cm and velocity < 30cm/sec) at a range of flows at the Mt. McKenzie section of the Gellibrand river, Victoria. (from Tunbridge and Glenane 1988).

DEFINING KEY FACTORS RELATING MARINE FISHES AND THEIR HABITATS

P.C. Young[1] and J.P. Glaister[2]

[1]*CSIRO Fisheries*
GPO Box 1538
Hobart TAS 7001

[2]*QDPI Fisheries*
GPO Box 46
Brisbane QLD 4001

Introduction

We have already heard about the key factors relating freshwater and estuarine fishes and their habitats and now we turn to the marine environment. The oceans are basins in the surface of the solid earth containing seawater. It is also evident that far more of the earth's surface is covered by sea (71%) than land (29%) and that perhaps we should refer to the "Ocean" rather than the Earth! Although the average depth of the oceans is close to 4 km (whilst the seas are generally about 1 km), relative to the horizontal dimensions of the oceans (5,000 to 15,000 km) this depth is small. An analogy for relative dimensions would be a sheet of lightweight typing paper (Marshall 1979). However there is a great degree of detail and structure in this relatively thin layer of the marine environment.

Broadly, the marine envrionment encompasses the coastal or continental margins, the ocean basins, the mid-oceanic ridge systems and the pelagic or ocean waters. Oceanographic conditions in the coastal areas differ markedly from those in the open sea. Some of the factors causing these differences include river runoff, tidal currents and the effects of shore boundaries on circulation (Pickard 1975). River runoff reduces salinity in surface waters, disperses suspended sediments and alters optical transparency. Tidal currents can cause large diurnal changes in the inshore volume of water and promote vertical mixing thereby reducing stratification. Thus strong currents across a rough bottom will, through turbulence, mix heated surface waters and bottom nutrients. Shore boundaries limit directions of current flow and allow deposition of sediment loads or scouring by wave action.

The continental shelf extends from the shore with an average gradient of 1 in 500. Its outer limit (the shelf-break) is set where the average gradient increases to about 1 in 20 to form the continental slope which continues to the deep-sea bottom. The Australian shelf occupies an area of approximately 2.6 million km^2 (or approximately one third of the land mass) and three quarters lies in the tropics. South of the Tropic of Capricorn, the shelf is generally less than 80 km wide, for instance the shelf width off New South Wales is less than 21 km. In the tropical region, the width exceeds 320 km in the north and 200 km in the west. The Australian shelf may be divided into eight regions, each characterised by its own typical bathymetry, areal development surface features, geological and biological structure. Thus the shelves of New South Wales and sub-tropical Queensland; Bass Strait and Tasmania; Southern Australia; sub-tropical Western Australia; North-West Australia; Arafura Sea; Gulf of Carpentaria; and the tropical shelf of Queensland, may each be separately distinguished.

The continental slope falls an average vertical depth of about 4 km from the shelf break to the deep-sea bottom but can drop as much as 9 km vertically over a relatively short distance. The material of the continental slope is predominantly mud, with some rock outcrops. The slope descends to depths near 2 km before merging with the gentler gradients (1:100 to 1:700) of the continental rise and this then grades to the deep-sea bottom. Over much of the oceans, the continental slopes and rises are covered with terrigenous or hemipelagic muds, mixtures of pelagic ooze, and debris derived from the continents. These muds are coloured red, green or black due to varying degrees of oxidation of the organic matter (Heezen and Rawson 1977). The proximity to land and hence the relatively rich shelf waters has much to do with the high content of organic matter in terrigenous muds, which in turn is the basis for the comparatively high productivity of benthic life over and on the continental margins.

The deep-sea bottom is the most extensive area, depths of 3 to 6 km being found over 76% of the ocean basins, with 1% of greater depth. The deepest parts of the oceans are referred to as trenches, and most occur in the Pacific Ocean. Down to about 5 km the ocean basins are covered with calcareous (derived from zooplankton) and siliceous (derived from phyto- and zooplankton) oozes and at depths greater than 5 km, with inorganic "red" clays.

Covering all of this is the ocean water, the pelagic environment. Seawater is a complex solution which contains most of the known elements. Chloride, sulphate, sodium, magnesium and potassium ions make up the principle dissolved constituents. The total amount of dissolved material in parts per thousand (ppt) of seawater is at a minimum (34.5) near the equator, rises to a maximum (35.8) at about latitude 20^0S and then decreases to 34.0 or less towards high latitudes. Though the salinity varies, the proportion of the principal ions is almost constant, due to continual mixing and circulation.

Radiant energy from the sun reaches the sea surface directly or as radiation scattered by the atmosphere. Some of this energy is reflected, the rest enters the sea surface, is absorbed and raises the temperature of the water. The outstanding thermal feature of the ocean is that there is an upper warm layer of water, mainly in tropical and sub-tropical latitudes, floating on a cold ocean. In fact, three quarters of the volume has temperatures between 0^0 and 6^0C and salinities between 34.6 and 34.8 ppt.

The density of seawater is an important property which allows oceanographers to trace the origins, equilibrium levels and movements of water masses. Density depends upon pressure as well as salinity and temperature and increases by 1 atmosphere (1 kg per cm^2) each increase in depth of 10 m. Small changes in temperature, salinity and density cause large scale mixing of water masses (Pickard 1963).

Evaporation, condensation and the release of latent heat cause atmospheric circulation (wind) that in turn cause waves and drive ocean currents. The wind-driven circulation of the ocean differs from the thermohaline circulation of the water masses (maintained by the pull of gravity on denser seawater types that sinks under less dense water and then flows along stratified levels of equilibrium). This wind-driven circulation consists of currents that flow horizontally in the upper few hundred metres of the ocean and as a consequence of the Coriolis Force, move in a direction to the left of the wind direction in the southern hemisphere (Marshall 1979).

Within this vast environment are the world's marine animal communities. Thorson (1971) has estimated that there are about 160,000 species of marine animals of which 98 per cent are benthic and 2 per cent pelagic. Only about 5 per cent of the total number of marine animals live in the deep sea, though the proportion of pelagic forms is perhaps greater in these depths than in the shallower areas. The deep and shallow seas are also alike, in that animal life is most diverse in the sub-tropical and tropical belts.

Whether considering the deepest ocean depths or the relatively shallow continental shelves, the key factors relating marine fishes and their habitats are defined by unique complexes of physical and biological factors which constitute the unique environment or ecological niche (Marshall 1979).

Pelagic species

Young (1992) noted that of the 3,000 species of marine fishes identified from Australian waters, the feeding ecology of less than 20 species has been studied in detail. However these results suggested that marine fish larvae in Australian waters have similar feeding strategies to their counterparts elsewhere in the world. For example, the larvae of *Trachurus* species (Jack Mackerels) eat cladocerans and larval euphausids off the coast of Tasmania when they are available (Young and Davis in press) and this has also been reported in the northern hemisphere (Arthur 1976; Sinyukova 1964). Jordan (1992) described the oceanographic variability of the east coast of Tasmania and its likely effects on jack mackerel (*Trachurus declivis*). Harris *et al.* (1991) suggested that in this region the local westerly wind stress is moving cold, nutrient rich water from the Antarctic coupled with the large-scale, oceanographic circulation, the East Australian Current, which moved warm, nutrient-poor water southwards to produce annual variability in oceanic conditions during La Niña years. They concluded that there was a general decline in nutrients and a resultant decline in algal biomass and large zooplankters (euphausiids and salps) at this time. Jordan (1992) speculated that these changes may have resulted in concomitant changes in the abundance and distribution of prey and predators of *Trachurus declivis* larvae and horizontal advection of eggs and larvae. Any or a combination of these, Jordan argued, may have accounted for the 95% reduction between years, of larval and egg density of *T. declivis*. The adult distribution is similarly controlled by water conditions; reduced nutrients lead to drastic reduction in biomass of larger zooplankton, especially krill which, if rare, do not swarm; jack mackerel do not school in commercial quantities and the fishery fails (Harris *et al.* 1992). Thus the water masses may define the marine habitat for jack mackerel and a similar situation appears to occur with southern bluefin tuna (*Tunnus maccoyii*).

Deepsea species

The deep-sea environment has already been alluded to, but what of the species that occur there? Probably, the most well-known Australian example is the orange roughy (*Hoplostethus atlanticus*) which is the target species for a recently developed deepwater trawl fishery in Australia and New Zealand, and more recently in the Rockall Trench off NW Europe. The orange roughy occurs in waters deeper than 1000m, feeds on luminescent prawns, squid and fish (Bulman and Koslow 1992), only produces less than one-tenth the number of eggs as many other species of commercial fish and is at least 20 years old at first spawning (Smith 1991).

Spawning occurs once a year in the Antipodes from late April to early August and in very dense aggregations. A principal winter spawning aggregation occurs off the east coast of Tasmania near St. Helens (Smith 1991). The eggs are large (2 mm diameter) and are fertilised in the water, after which they drift up towards the surface and remain planktonic until they hatch (Robertson 1991). Duration of the egg stage and further larval development is unknown.

Orange roughy are found in the South Pacific with a discrete community of other deep sea fish species which co-occur at similar depths. Koslow *et al.* (in press) have shown that similar species are found in the deeper parts of the North Atlantic, but not in the North Pacific. They suggested that this discontinuous distribution is due to differences in the distribution of the Antarctic Intermediate water mass in which they live. This has a significant flux into the North Atlantic but not the North Pacific. Koslow *et al.* (in press) postulate that the absence of orange roughy and its associated fish community from the North Pacific is due to their distribution being determined by that of the physical circulation of Antarctic Intermediate water.

Another deepwater species, blue grenadier (*Macruronus novaezelandiae*), the most abundant commercial finfish species in the New Zealand deepwater fishery, is a serial spawner, releasing multiple batches of eggs in a season (Sullivan 1991). They are highly fecund, with an average sized female of 90 cm and spawn over 1 million eggs in a season (Hurst *et al.* in Sullivan 1991). In New Zealand, grenadier spawn in winter near canyon features. They form dense aggregations which disperse off the bottom at night when spawning occurs, thus positioning the released eggs in mid-water. The planktonic eggs and larvae are dispersed by currents. Growth is rapid, with juveniles reaching 27 cm by the end of the first year and mature adults of 60-70 cm by 4-5 years. Maximum age is thought to be 20-25 years.

However Bulman and Blaber (1986) and Gunn *et al.* (1989) have shown that off Tasmania, blue grenadier undergo a diurnal vertical migration to feed on mesopelagic prey, during the night. These themselves undergo a vertical migration in which they move to the surface waters during the day, and return to the midwater at night where they form the prey of blue grenadier (Young and Blaber 1986).

Shallow marine species

Hatcher *et al.* (1989) provide an excellent review of the principal research relevant to the conservation of tropical marine ecosystems (coral reef, mangroves, seagrasses). They provide a comparison (Table 1, pp 342-4) of the ways in which shallow, tropical marine ecosystems differ from their temperate counterparts at comparable depths. These include physico-chemical (temperature, light, dissolved oxygen, salinity, sediment), community structure (species diversity, biomass, population dynamics, eggs and larvae, and genetic variability) and biological functions (metabolism, reproduction, feeding and specialisation). Thus latitudinal differences (high species diversity, lower individual species biomass in the tropics) present particularly complex fishery management problems with multispecies interactions (Sainsbury 1982).

A high proportion of tropical organisms have an extended pelagic phase in their life histories in which they can only be identified as occupying generally similar areas to adults on parts of the continental shelf and adjacent slope (Young *et al.* 1986) and thus events relatively remote from their adult benthic or epibenthic habitat may influence survival of larval organisms in the oceanic environment. Alterations to oceanic circulation patterns or water quality, for example, could have dramatic impacts on recruitment and hence adult populations (Hatcher *et al.* 1989).

(a) Coral reef species

Coral reef ecosystems have been described as the oldest, most diverse and complex ecosystems on the planet (Davies and Montaggioni 1985). Much has been written describing pattern within these ecosystems (e.g. Williams 1982; Bradbury and Young 1981), but it is the understanding of processes and their quantification, which are likely to provide insights needed for management (Kenchington 1990). At the core is the understanding of processes operating at different spatial scales and at different temporal scales. For example, the role of pelagic life stages and recruitment-controlling processes in population and community dynamics of fish (Sale 1980), the role of local disturbance and larger scale natural disasters and catastrophes on coral reef community structure and whether processes are operating at the coral head, patch reef, reef platform or reef province spatial scale (Doherty 1987).

In addition, a range of anthropogenic (human origin) effects including water pollution (sewage, industrial and agricultural runoff, thermal effluent); turbidity and sedimentation (land clearing, dredging, construction and mining); coral removal (mining, dredging, collecting, vessel grounding, anchoring) and removal of living organisms (fishing, collecting, inadvertent removal of non-targeted species), combine to alter available habitat. Coral reef fishes typically are highly fecund, can be serial spawners and have a long breeding season. They have a pelagic larval stage and sometimes the egg is also pelagic. Thus although the adults may be relatively sedentary, the offspring are highly mobile through dispersal by oceanic currents (Sale 1980).

The mechanisms by which larval fishes return to the reef area is uncertain. Most researchers have assumed that the trip must be largely passive (Leis 1977; Sale 1971) because larvae are unable to swim against any but the weakest currents. However, they seem unable to detect cues to the presence of a reef until they are quite close (Sale 1980). Gyres in the lee of islands may concentrate larvae, thereby preventing their loss (Leis and Miller 1976). Because of this relationship of larval distribution to this and other patterns of water circulation, Johannes (1978) considered seasonal patterns in spawning have arisen as responses to patterns of water circulation. By contrast with this passive dispersion as larvae, once fish return to the reef there is usually a marked change in their habit in which they become relatively strongly site-attached and sedentary for the remainder of their life (Doherty 1987).

Once settled, areas of more or less homogenous habitat on a reef will contain an assemblage of fishes with certain characteristics. The species richness of such assemblages will largely be a function of its numerical size (unless it is large enough to include all species available). Similarly the species comprising the assemblage may be expected to exhibit a wide range of food and microhabitat requirements, and display complex positive or negative behavioural interactions with one another. These could be trophic, competitive or mutualistic relationships and define social structures. A large number of factors could then determine one or more features of the observed structure. These could include limiting resources (food, living space) predation and other disturbances, and recruitment (Sale 1980).

The availability of topographically-structured living space may limit populations of reef fish (Bradbury and Young 1981). The common structural types of coral reef include *barrier, fringing* and *platform* reefs. A generalised outer or barrier reef would include (after Veron 1986): *lower slope* (limited coral, usually extensive colonies of one species or genus, light limited and influenced strongly by water clarity, currents and the steepness of the slope); *upper slope* (maximum coral species diversity at 20 m, mixed community not dominated by any one species or limited by light, rich structure); *reef front* (narrow zone, exposed at low tide, few short stocky corals able to withstand the force of the ocean swell); *outer reef flats* (hammered by surge, highest point of reef and least populated by corals); *inner reef flats* (differs from former in having much loose rubble and only partly consolidated substrate, sand and rubble intermixed with solid reef rock covered with few coral species); *lagoons* (rocky substrates of inner reef flats eroded leaving a sandy floor, surrounded by reef, good circulation via tidal currents, relatively few corals but, as protected from wave action, may have elaborate forms); *back reef slopes* (less regular than outer reef slopes, with overhangs, caves and deep canyons, coral lush on gentle slopes but bare on steep slopes); and *inner-reef sea floor* (rubble, sand or mud, few corals but rich community of other species, subject to terrestrial influences).

A typical fringing reef would be found around high islands or along the mainland foreshore and generally have: *inner intertidal mudflats* fringed with extensive mangrove swamps (tidally flushed, subject to storms and terrestrial runoff, little structural complexity); *outer intertidal mudflats* (increasing depth seaward, increasing coral cover in the form of microatolls, diversity increasing with depth); and the *outer slope* (the top exposed only at very low tides, broad bank of massive coral colonies and species diversity heavily influenced by turbidity (Veron 1986).

In addition, solitary platform reefs, corals on rocky shores, algal communities, caves and deep water communities also occur, providing extensive and diverse marine habitat for fish. As with barrier and fringing types, the physical environment (light, wave action, sedimentation, salinity, tidal range, food and inorganic nutrients, temperature and bathymetry) largely controls community structure. Such structure largely determines marine fish habitat.

(b) Mangrove species

Mangrove communities are marine tidal forests, taxonomically diverse, with structure characterised by adaptation to differences in temperature, soil salinity, frequency of tidal inundation, sedimentation, soil chemistry, degree and frequency of freshwater flow and ground water availability (Hatcher *et al.* 1989). The role of mangroves includes the stabilisation of sediments deposited by physical forces (Robertson and Duke 1987; Blaber *et al.* 1989) and the provision of food, space and shelter to a range of mangrove-dependent biota (Milward 1979). The species of fish that rely on mangrove habitats include permanent residents, intermittent visitors and seasonal transients (as eggs, larvae, juveniles or adults). Species occurrence varies geographically, and also in relation to local topography, tidal range, salinity variation and turbidity (Saenger *et al.* 1977).

The importance of mangroves to fish has long been a truism amongst ecologists (Boesch and Turner 1984) and has been demonstrated by Robertson and Duke (1987), who sampled juvenile fish communities in mangrove, seagrass and mudflat communities in north-east tropical Australia. They showed that the densities of fish and prawns in mangrove habitats are an order of magnitude greater than in the adjacent nearshore habitats. Blaber *et al.* (1989) and Robertson and Duke (1990) similarly demonstrated markedly higher mean biomasses of fishes from mangrove creeks, inlets and intertidal mudflats adjacent to mangroves, than from surrounding areas. It is unlikely that trophic relationships alone could account for these differences. Other factors could include greater available living spaces, the exclusion of marine predators via the inflow of freshwater (Blaber *et al.* 1985), the physical barrier to predators provided by mangrove roots (Robertson 1988) and sheltering effects of high turbidity levels (Blaber and Blaber 1980).

(c) Seagrass species

Seagrass communities are found from the tropics to the sub-polar regions and along with coral reefs and mangrove forests, rank among the most productive recorded for natural communities (Odum 1982). As the structural complexity of coastal ecosystems has been revealed by research, the importance of seagrass has become increasingly recognised (Larkin *et al.* 1989). Seagrasses provide food and shelter for a variety of plant and animal species, increase the primary productivity of coastal waters, recycle nutrients, stabilise sediments and provide nursery habitat for a range of fish and crustacea (among other groups). Seagrass communities also often occur in proximity to urban (or potentially urban) areas and anthropogenic events can cause pronounced destruction through reclamation, dredging, modified hydrological regimes (wave action), overgrazing, siltation, eutrophication, mechanical damage (worm digging), point source pollution and modified tidal regimes (channelling). Natural disasters (cyclones, floods) and cycles (closing of coastal lagoons, dugong grazing) are also significant (West 1989).

Like mangroves, seagrasses physically stabilise sediments since leaves and roots provide a physical baffle to water flows, allowing particles (and larvae) to settle (Hatcher *et al.* 1989).

The roots and rhizomes also matt together, binding the particles and minimizing erosion. Seagrasses in coarser-particle sediments tend to have a greater root mass, presumably for greater absorption of nutrients in the nutrient-poor, coarser sediments (Wood and Johannes 1975). Seagrasses also allow chemical exchange to the sediments and provide a substrate for algal epiphytes.

Again like mangroves, seagrasses provide shelter and habitat for small fish and juvenile prawns (Young and Wadley 1979; Coles *et al.* 1987). In the Caribbean, predation in seagrass communities occurs at night when predators leave the shelter of nearby coral reefs to forage over seagrass beds (Phillips and McRoy 1980). In tropical Australia seagrass meadows dominated by *Halophila* and *Halodule* are important as nursery grounds for several commercially important penaeid prawns (Coles and Lee Long 1985; Staples *et al.* 1985; Coles *et al.* 1987; Poiner 1980; Poiner *et al.* 1987).

Many species present in seagrass nursery areas as juveniles may move to other habitats upon reaching a critical length at which the seagrass may no longer provide adequate shelter (Minello and Zimmerman 1983; Pollard 1984). In their review of the literature Orth *et al.* (1984) identified a number of factors to explain the high densities of some species in a range of seagrass beds and in particular, the role of plant architecture in the vulnerability of prey species to predation. They further proposed that predation and competition could account for reported differences in seagrass structure and associated faunal abundance in some of the work they reviewed. Heck and Thoman (1981) examined how the success rate of predators varied with the density of submerged vegetation both experimentally and in the field. They found that high densities of artificial eelgrass hindered the success of predators of grass shrimp. They also found (1984) as did Heck *et al.* (1989) that vegetated bottoms supported higher numbers of decapods than unvegetated areas. Capehart and Hackney (1989) identified similar differences in salt-marsh habitat.

Werner *et al.* (1983) investigated foraging costs (through an application of optimal foraging theory) and found that fish could maximise feeding activity through changing habitats. Weinstein and Heck (1979) for example, suggest a major functional difference between tropical and temperate seagrass habitats: coral reefs adjacent to tropical seagrass beds provide shelter and physical complexity not found in temperate beds. However physical barriers (islands, water circulation) may limit this generalisation (Sogard *et al.* 1989).

However, comparatively few animals consume seagrasses directly. The ones that do, include green turtles (Limpus and Reed 1984), dugong (Poiner *et al.* 1987) and some sea urchins, surgeonfishes and parrotfishes (Kirkman and Young 1981; Hatcher *et al.* 1989). Seagrasses also provide a substrate for a variety of epiphytic algae. Orth *et al.* (1984) estimated production of algal epiphytes attached to seagrass blades approaches 20% of the seagrass production and that epiphytes are more important as food for associated fauna than are the more refractory seagrass blades.

Seagrass communities generally have been extensively studied worldwide yet such habitats in tropical and temperate Australia have been comparatively ignored. Gulf of Carpentaria habitats (Coles and Lee Long 1985; Poiner *et al.* 1987), East Coast of Queensland (Coles *et al.* 1985; Young and Kirkman 1975; Kirkman 1978; Poiner 1984; Lennon and Luck 1990), New South Wales and Victoria (Larkin *et al.* 1989), South Australia (Neverauskas 1987) and Western Australia (Silberstein *et al.* 1986) have been investigated and more recently the use of satellite imagery and Geographical Information Systems (GIS) have enhanced inventory surveys. Most such studies have been mainly concerned with establishing the importance of wetland areas to commercial and other fish and prawn species.

Key factors—A summary

This review of key factors relating marine fishes and their habitats has ranged from the deepest to the shallowest marine waters and their habitats, and through the full range of life history forms from egg to adult. But what *is* fish habitat? Most simply, it is where they live. Such areas can be classified, for example using Fry's system of habitat classification, and then through the limiting factors and controlling factors defined for each habitat type. These factors are physical and biological and together complexes of these factors constitute unique fish habitats.

In this brief overview of a huge topic, we have attempted to highlight the principal marine fish habitats and some of the important factors that define them. It is worth noting that there is a gradation in our knowledge base from shallowest to deepest. In other words, because it is easier (and cheaper!) to study a seagrass bed in a few metres than the benthos in a few kilometres, we *know* more about the shallower habitats. But that bias aside, some generalisations can perhaps be made regarding marine fish habitat.

We have seen that structural complexity declines from shallow coral reef/mangrove forest/seagrass bed to deep sea benthos. So too does comparative productivity. We know that oceanic circulation (wind-driven and thermohaline-driven) becomes increasingly important with depth in defining fish habitat. We have seen the numbers of species decline with depth and from lower to higher latitudes. The impacts of the terrestrial-ocean interface events (runoff, topography and circulation and tides) decreases from shallower to deeper waters. Finally, natural disasters, catastrophes and anthropogenic impacts are probably of more importance in shallower than deeper waters.

It seems that, in terms of the topic of this workshop, and in providing managers with what they need, when relating marine fishes and their habitats we need to carefully define the question or risk being lost in complexity.

What scale or level of complexity is needed? What are the important factors limiting and controlling? The present threats, given earlier presentations, certainly point to a continuum from most threatened (freshwater) habitats to least threatened (deep sea) habitats.

References

Arthur, D.K. (1976). Food and feeding of larvae of three fishes occurring in the California current, *Sardinops sagax*, *Engraulis mordax*, and *Trachurus symmetricus*. *Fishery Bulletin, U.S.* **74**, 517–530.

Blaber, S.J.M. and T.G. Blaber (1980). Factors affecting the distribution of juvenile estuarine and inshore fishes. *Journal of Fish Biology* **17**, 143–62.

Blaber, S.J.M., J.W. Young and M.C. Dunning (1985). Community structure and zoogeographic affinities of the coastal fishes of the Dampier region of north-western Australia. *Australian Journal of Marine and Freshwater Research* **36**, 247–66.

Blaber, S.J.M., D.J. Brewer and J.P. Salini (1989). Species composition and biomasses of fishes in different habitats of a tropical northern Australian estuary: their occurrence in the adjoining sea and estuarine dependence. *Estuarine, Coastal and Shelf Science* **29**, 509–31.

Boesch, D.F. and R.E. Turner (1984). Dependence of fishery species on salt marshes: the role of food and refuge. *Estuaries* **7**, 460–8.

Bradbury, R.M. and P.C. Young (1981). The effects of a major forcing function, wave energy, on a coral reef ecosystem. *Marine Ecology Progress Series* **5**, 229–241.

Bulman, C.M. and S.J.M. Blaber (1986). Feeding ecology of *Macruronus novaezelandae* (Hector) (Teleostei: Merlucciidae) in South-eastern Australia. *Australian Journal of Marine and Freshwater Research* **37**, 621–39.

Bulman, C.M. and J.A. Koslow (1992). Diet and food consumption of a deep sea fish, orange roughy *Hoplostethus Atlanticus* (Pisces: Trachichthyidae) off South-Eastern Australia. *Marine Ecology Progress Series* **82**, 115–129.

Capehart, A.A. and C.T. Hackney (1989). The potential role of roots and rhizomes in structuring salt-marsh benthic communities. *Estuaries* **12(2)**, 119–22.

Coles, R.G. and W.J. Lee Long (1985). Juvenile prawn biology and the distribution of seagrass prawn nursery grounds in the south-eastern Gulf of Carpentaria. *In*: P.C. Rothlisberg, B.J. Hill and D.J. Staples (Eds.)(1985). "Second Australian National Prawn Seminar", NPSZ, CSIRO, Cleveland, Australia, pp.55–60.

Coles, R.G., W.J. Lee Long and L.C. Squire (1985). Seagrass beds and prawn nursery grounds between Cape York and Cairns. QDPI Information Series, QI 85017.

Coles, R.G., W.J. Lee Long, B.A. Squire, L.C. Squire and J.M. Bibby (1987). Distribution of seagrasses and associated juvenile commercial penaeid prawns in north-eastern Queensland waters. *Australian Journal of Marine and Freshwater Research* **38**, 103–19.

Davies, P.J. and I. Montaggioni (1985). *In Proceedings of the Fifth International Coral Reef Congress* (Delesalle, *et al.* (eds), Antenne Museum -EPHE, Moorea **3**, 95–102.

Doherty, P.J. (1987). Light-traps: useful but selective, devices for quantifying the relative abundance of larval fishes. *Bulletin of Marine Science* **41**, 423–31.

Gunn, J.S., B.D. Bruce, D.M. Furlani, R.E. Thresher and S.J.M. Blaber (1989). Timing and location of spawning of Blue Grenadier, *Macruronus novaezelandiae* (Teleosti: Merlucciidae), in Australian coastal waters. *Australian Journal of Marine and Freshwater Research* **40**, 97–112.

Harris, G.P., F.B. Griffiths, L.A. Clementson, V. Lyne and H. Van Der Doe (1991). Seasonal and interannual variability in physical processes, nutrient cycling and the structure of the food chain in Tasmania shelf waters. *Journal of Plankton Research* **13**, 109–131.

Harris, G.P., F.B. Griffiths and L.A. Clementson (1992). Climate and the fisheries off Tasmania - interactions of physics, food chains and fish *in* Payne A.S.L., K.H. Brink, K.H. Mann and R. Hillborn (eds) (1992) "Benguela Trophic Functioning" *South African Journal of Marine Science* **12**, 585–597.

Hatcher, B.G., R.E. Johannes and A.I. Robertson (1989). Review of research relevant to the conservation of shallow tropical marine ecosystems. *Oceanography and Marine Biology Annual Review, 1989,* **27**, 337–414, Margaret Barnes (ed.), Aberdeen University Press.

Heck, K.L. and T.A. Thoman (1981). Experiments on predator-prey habitats in vegetated aquatic habitats. *Journal of Experimental Marine Biology and Ecology* **53**, 125–34.

Heck, K.L. and T.A. Thoman (1984). The nursery role of seagrass meadows in the upper and lower reaches of the Chesapeake Bay. *Estuaries* **7(1)**, 70–92.

Heck, K.L., K.W. Abel, M.P. Fahay and C.J. Roman (1989). Fishes and decapod crustaceans of Cape Cod eelgrass meadows: species composition, seasonal abundance patterns and comparisons with unvegetated substrates. *Estuaries* **12(2)**, 59–65.

Heezen, B.C. and M. Rawson (1977). Influence of abyssal circulation on sedimentary accumulations in space and time. *Marine Geology* **23**, 173–196.

Johannes, R.E. (1978). Reproductive strategies of coastal marine fishes in the tropics. *Environmental Biology of Fishes* **3**, 65–84.

Jordan, A.J. (1992). Interannual variability in the oceanography of the east coast of Tasmania and its effects on Jack Mackerel (*Trachurus declivis*) larvae. *In* Hancock, D.A. (ed.)(1992). "Larval Biology," *Australian Society for Fish Biology Workshop, Hobart, 20 August 1991, Bureau of Rural Resources, Proceedings No.15*, AGPS, Canberra, 211pp.

Kenchington, R.A. (1990). "Managing Marine Environments," Taylor and Francis, New York, 248pp.

Kirkman, H. (1978). Decline of seagrass in northern areas of Moreton Bay. *Aquatic Botany* **5**, 63–76.

Kirkman, H. and P.C. Young (1981). Measurement of heath and echinoderm grazing on *Posidonia oceanica* (L) Delile. *Aquatic Botany* **10**, 329–338.

Koslow, J.A., C.M. Bulman and J.M. Lyle (in press). The midslope demersal fish community off South-eastern Australia. *Deep Sea Research.*

Larkum, A.W.D., A.J. McComb and S.A. Shepherd (1989). "Biology of Seagrasses: a treatise on the biology of seagrasses with special reference to the Australian region," Elsevier, Amsterdam.

Leis, J.M. (1977). Systematics and zoogeography of the porcupine fishes (Diodon, Diodontidae, Tetradontiformes), with comments on egg and larval development. *Fisheries Bulletin, US* **76**, 535–567.

Leis, J.M. and J.M. Miller (1976). Offshore distributional patterns of Hawaiian fish larvae. *Marine Biology* **36**, 359–67.

Lennon, P. and P. Luck (1990). Seagrass mapping using LANDSAT TM Data: a case study in southern Queensland. *Asian-Pacific Remote Sensing Journal* **2**, 1–7.

Limpus, C.J. and P.C. Reed (1984). The green turtle *Chelonia midas* in Queensland. Population structure in a coral reef feeding ground. *In* G. Grigg, R. Shine and H. Ehmann (Eds.) "The Biology of Australasian Frogs and Reptiles," Surrey Beatty and Sons, Australia, pp. 53–61.

Marshall, N.B. (1979). "Developments in Deep-Sea Biology," Blandford Press, Dorset, 566pp.

Milward, N.E. (1979). Mangrove-dependent biota. In B.F. Clough (ed.). "Mangrove Ecosystems in Australia, Structure, Function and Management, "*Proceedings of the Australian National Mangrove Workshop*, AIMS, April 1979, Books Australia, 302pp.

Minello, T.J. and R.J. Zimmerman (1983). Fish predation on juvenile brown shrimp, *Penaeus aztecus* Ives: the effect of simulated Spartina structure on predation rates. *Journal of Experimental Marine Biology and Ecology* **72**, 211–31.

Neverauskas, V.P. (1987). Accumulation of periphyton biomass on artifical substrates deployed near a sewage sludge outfall in South Australia. *Estuarine and Coastal Shelf Science* **25**, 509–17.

Odum, W.E. (1982). Environmental degradation and the tyranny of small decisions. *Bioscience* **32**, 728.

Orth, R.S., K.L. Heck and J. van Montrans (1984). Faunal communities in seagrass beds: a review of the influence of plant structure and prey characteristics on predator-prey relationships. *Estuaries* **7(4)**, 339–50.

Phillips, R.C. and C.P. McRoy (Eds.)(1980). "Handbook of Seagrass Biology: and Ecosystem Approach," Garland STPM Press, New York, pp.173–98.

Pickard, G.L. (1963). "Descriptive Physical Oceanography," Permagon Press, Oxford, 200pp.

Pickard, G.L. (1975). "Descriptive Physical Oceanography: an Introduction," Permagon Press, Oxford, 2nd Edition, 214pp.

Poiner, I.R. (1980). A comparison between species diversity and community flux rates in the macrobenthos of an infaunal sand community and a seagrass community of Moreton Bay, Queensland. *Proceedings of the Royal Society of Queensland* **91**, 21–36.

Poiner, I.R. (1984). Interspecific interactions: their role in structuring multispecific seagrass communities. *AES Monograph* **1**, 55–82.

Poiner, I.R., D.J. Staples and R. Kenyon (1987). Seagrass communities of the Gulf of Carpentaria, Australia. *Australian Journal of Marine and Freshwater Research* **38**, 121–31.

Pollard, D.A. (1984). Review of ecological studies on seagrass - fish communities, with particular reference to recent studies in Australia. *Aquatic Botany* **18**, 3–42.

Robertson, A.I. (1988). Decomposition of mangrove leaf litter in tropical Australia. *Journal of Experimental Marine Biology and Ecology* **116**, 235–47.

Robertson, A.I. and N.C. Duke (1987). Mangroves as nursery sites: comparisons of the abundance and species composition of fish and crustaceans in mangroves and other nearshore habitats in tropical Australia. *Marine Biology* **96**, 193–205.

Robertson, A.I. and N.C. Duke (1990). Mangrove fish-communities in tropical Queensland, Australia: spatial and temporal patterns in densities, biomass and community structure. *Marine Biology* **104(3)**, 369–79.

Robertson, D.A. (1991). The New Zealand Orange Roughy Fishery: an overview. In Abel, K., M. Williams and P. Smith (Eds.)(1991). "Australian and New Zealand Southern Trawl Fisheries Conference: Issues and Opportunities," *Bureau of Rural Resources, Proceedings No.10*, Pirie Printers Sales, Canberra, 266pp.

Saenger, P., M.M. Specht, R.L. Specht and V.J. Chapman (1977). Mangal and coastal salt-marsh communities in Australasia. *In* V.J. Chapman (ed.). "Ecosystems of the World I. Wet Coastal Ecosystems," Elsevier, Amsterdam, pp293–345.

Sainsbury, K.J. (1982). The ecological basis of tropical fisheries management. In Pauly, D. and G.J. Murphy (Eds.) "Theory and Management of Tropical Fisheries, pp. 167–78, ICLARM Conference Proceedings **9**.

Sale, P.F. (1971). The reproductive behaviour of the pomacentrid fish *Chromis caeruleus*. *Z. Tierpsychol.* **29**, 156–64.

Sale, P.F. (1980). The ecology of fishes on coral reefs. *Oceanography and Marine Biology Annual Review, 1980* **18**, 367–421, Harold Barnes (ed.), Aberdeen University Press.

Silberstein, K., A.W. Chiffings and A.J. McComb (1986). The loss of seagrass in Cockburn Sound, Western Australia. III. The effects of epiphytes on productivity of *Posidonia australis* Hook. *Aquatic Botany* **24**, 355–71.

Sinyukova, V.I. (1964). The feeding of Black Sea horsemackerel larvae. *Trudy Sevastopolbskoi Biologiche Skoi Stansii*, pp. 302–325.

Smith, A.D.M. (1991). As many red fish as live in the sea: Orange Roughy in Australia. *In* Abel, K., M. Williams and P. Smith (Eds.)(1991). "Australian and New Zealand Southern Trawl Conference: Issues and Opportunities, *Bureau of Rural Resources, Proceedings No.10*, Pirie Printers Sales, Canberra, 266pp.

Sogard, S.M., G.V.N. Powell and J.G. Holmquist (1989). Utilisation by fishes of shallow, seagrass-covered banks in Florida Bay: 1. Species composition and spatial heterogeneity. *Environmental Biology of Fishes* **24(1)**, 53–65.

Staples, D.J., D.J. Vance and D.S. Heales (1985). Habitat requirements of juvenile penaeid prawns and their relationship to offshore fisheries. *In*: P.C. Rothlisberg, B.J. Hill and D.J. Staples (Eds.)(1985). "Second Australian National Prawn Seminar," NPSZ, CSIRO, Clevaland, Australia, pp.47–54.

Sillivan, K.J. (1991). A review of the Hoki fishery and research on Hoki stocks in New Zealand. In Abel, K., M. Williams and P. Smith (Eds.)(1991). "Australian and New Zealand Southern Trawl Fisheries Conference: Issues and Opportunities", *Bureau of Rural Resources, Proceedings No.10*, Pirie Printers Sales, Canberra, 266pp.

Thorson, G. (1971). "Life in the Sea," Weidenfeld and Nicolson, London, 256pp.

Veron, J.E.N. (1986). "Corals of Australia and the Indo-Pacific," Angus and Robertson, Hong Kong, 644pp.

Weinstein, M.P. and K.L. Heck Jnr (1979). Ichthyofauna of seagrass meadows along the Caribbean coast of Panama and in the Gulf of Mexico: composition, structure and community ecology. *Marine Biology* **50**, 97–107.

Werner, E.E., G.G. Mittelbach, D.J. Hall and J.F. Gilliam (1983). Experimental tests of optimal habitat use in fish: the role of relative habitat profitability. *Ecology* **64(6)**, 1525–39.

West, R. (1989). How are seagrass meadows destroyed and when should attempts be made to restore them? *In* G. Edgar and H. Kirkman (eds.)(1989). "Recovery and Restoration of Seagrass Habitat of Significance to Commercial Fisheries," *Victorian Institute of Marine Sciences, Working Paper (19)*, 37pp.

Williams, D. McB. (1982). Patterns in the distribution in fish communities across the central Great Barrier Reef. *Coral Reefs* **1**, 35–43.

Wood, E.J.F. and R.E. Johannes (Eds.)(1975). "Tropical Marine Pollution," Elsevier, Amsterdam, pp. 64–74.

Young, P.C. and H. Kirkman (1975). The seagrass communities of Moreton Bay, Queensland. *Aquatic Botany* **1**, 191–202.

Young, P.C. and V. Wadley (1979). Distribution of Shallow-Water Epibenthic Macrofauna in Moreton Bay, Queensland, Australia. *Marine Biology* **53**, 83–97.

Young, P.C., J.M. Leis and H. Hausfeld (1986). Seasonal and spatial distribution of fish larvae in waters over the North West Continental Shelf of Western Australia. *Marine Ecological Progress Series* **31**, 209–22.

Young, J.W. (1992). Feeding ecology of marine fish larvae: an Australian perspective In Hancock, D.A. (ed.)(1992). "Larval Biology," *Australian Society for Fish Biology Workshop, Hobart, 20 August 1991, Bureau of Rural Resources, Proceedings No.15*, AGPS, Canberra, 211pp.

Young, J.W. and S.J.M. Blaber (1986). The feeding ecology of three species of midwater fish associated with the continental slope off Tasmania. *Marine Biology* **98**, 147–156.

Young, J.W. and T.L.O. Davis (in press). Feeding ecology of larvae of jack mackerel, *Trachurus declivis* (Pisces: Carangidae), from coastal waters of Eastern Tasmania. *Marine Biology*.

DEFINING KEY FACTORS RELATING FISH POPULATIONS IN ESTUARIES AND THEIR HABITATS

N.R. Loneragan

CSIRO Division of Fisheries, Cleveland Marine Laboratories
PO Box 120
Cleveland QLD 4163

Summary

Some of the recent literature is reviewed and results of detailed studies of fish and crustacean populations in temperate estuaries of south-western Australia and New South Wales are synthesised and the approaches to these studies discussed. Studies on the west coast of Australia have concentrated on defining seasonal, annual and spatial patterns of change in the fish fauna of the Swan and Peel-Harvey estuaries. The emphasis has been on obtaining detailed knowledge of the life history strategies of fish in estuaries and interpreting the main factors affecting the fish populations and community structure in light of this information. On the east coast, more effort has been directed towards evaluating the importance of various habitats to fish in estuaries, particularly seagrass habitats in several different estuarine and inshore coastal systems. Conventional sampling techniques (i.e. beach seines, gill nets, otter and beam trawls) have been used to study fish populations in estuaries of both regions. In addition, in Western Australia, commercial catch data in the Peel-Harvey and Swan estuaries have been used to assess how fish populations have responded to the marked eutrophication in the former system. Artificial seagrass has also been used in NSW to test hypotheses about the importance of seagrass to larval and juvenile fish.

Introduction

Estuaries are complex ecosystems in which environmental conditions are influenced by water derived from both riverine and marine sources. Furthermore, a wide variety of species of fish, representing different life cycle categories, is found in estuaries (Loneragan et al. 1989; Potter et al. 1990). The influence of environmental factors can vary greatly among the different life cycle categories of fish. For example, marine species are generally influenced to a greater extent by changes in salinity than estuarine or anadromous species.

In this paper, I outline some of the environmental variation found in estuaries, discuss the fish fauna found with estuarine systems and examine approaches to the study of key factors which influence the fish fauna. Most of the discussion will focus on temperate estuarine systems in Australia, particularly those in south-western and south-eastern Australia. In reviewing studies of estuarine fish populations, it is important to consider details of the spatial and temporal scales that were used, as these can greatly influence the results and interpretation of the important factors influencing the fish populations.

The estuarine environment

Pritchard (1967) proposed the first widely accepted definition of an estuary, namely that an estuary is a semi-enclosed body of water which has a free connection with the open sea and within which sea water is measurably diluted with fresh water derived from land drainage. However, this definition does not cover the types of environment found in many estuaries in southern Africa and parts of Australia (particularly the south-west) where sand bars may form at the mouths of estuaries and the water within the estuary can become markedly hypersaline. In order to include these systems, Day (1980; 1981) proposed that an estuary is a partially enclosed coastal body of water which is either permanently or periodically open to the sea and within which there is a measurable variation of salinity due to the mixture of sea water with fresh water derived from land drainage.

Since estuaries are regions where there is a mixing of water derived from both oceanic and riverine sources, they are typically environments where characteristics of the water column can undergo pronounced fluctuations (Day et al. 1981; Haedrich 1983). In addition to a salinity gradient along the estuary, there may be abrupt and large changes in salinity, temperature, dissolved oxygen, turbidity, flow and nutrient levels (Haedrich 1983; Cloern and Nichols 1985). The amount of fresh water flowing into estuarine systems varies greatly according to latitude, the seasonal pattern and quantity of rainfall, and characteristics of the catchment (e.g. amount of clearing, soil type and gradient of slope into the tributary rivers) (Day 1981). An example of the variation in patterns of annual rainfall is provided below. In the temperate lowlands of Europe, the annual rainfall is approximately 650-760 mm and it rains throughout the year. Relatively continuous rainfall also occurs in Melbourne, south-eastern Australia and Christchurch, New Zealand (Day 1981). By contrast, both Cape Town in South Africa and Perth in south-western Australia have a similar total annual rainfall, but most rain falls in winter (Day 1981; McComb et al. 1981). In tropical regions of Australia, virtually all the rainfall is in summer and early autumn, and the total annual rainfall can be at least twice that of the above temperate regions (e.g. Davis 1988). The pattern of rainfall influences the timing of changes in hydrological conditions within estuarine systems and hence the times when these environments are suitable for colonisation by marine flora and fauna.

The influence of marine waters on the hydrology of an estuary can also vary greatly between systems. Thus, in funnel shaped estuaries, i.e. estuaries with a broad mouth and a channel width which decreases progressively with distance upstream, the influence of tides can be greatly accentuated (Day 1981). Contrasting with this pattern of tidal accentuation, is the situation in many estuaries, particularly those in southern Australia, where the tidal energy and the resulting change in water level, diminishes with distance from the estuary mouth (Day 1981).

The fish fauna of estuaries

Relatively large numbers of the juveniles of some marine teleosts regularly enter estuaries in the temperate regions of both the Northern and Southern hemispheres (e.g. Gunter 1938; Pearcy and Richards 1962; McHugh 1967; Cronin and Mansueti 1971; Day et al. 1981; Haedrich 1983; Claridge et al. 1986). It is for this reason that estuaries have often been referred to as fish nursery areas (Cronin and Mansueti 1971; Haedrich 1983; Blaber 1985; Potter et al. 1990). Many other marine species are found only in small numbers in estuaries and generally in the high salinity regions towards their mouths (Haedrich 1983; Claridge et al. 1986).

Estuaries are used by anadromous species as a route between feeding areas in marine waters and their spawning grounds in freshwater, whereas they allow the reverse migration in the case of catadromous species (Day et al.

1981; Haedrich 1983; Dando 1984). A few species of teleost, almost certainly of marine origin, have evolved the ability to complete the whole of their life cycle within estuarine environments (Ross and Epperley 1985; Potter *et al.* 1986a; 1990). Some of these 'estuarine' species are also represented by populations in marine environments (Lenanton 1977; Prince *et al.* 1982; Chrystal *et al.* 1985; Potter *et al.* 1986b). The upper reaches of estuaries are occasionally penetrated by a few of the more euryhaline freshwater teleosts (Cronin and Mansueti 1971; Day *et al.* 1981).

The very high numbers of the juveniles of some of the marine species of teleost found in estuaries have led to these species frequently being included with estuarine and diadromous species in a category termed estuarine-dependent (Clarke *et al.* 1969; Pollard 1976; 1981; Van den Broek 1979; Fourtier and Legget 1982; Haedrich 1983; Beckley 1984; Blaber 1987). Since several recent studies have shown that a number of these marine species which are found in abundance in estuaries also utilize marine environments extensively at the same stage of their life cycles (Lenanton 1982; Beckley 1984; Smale 1984; Lenanton and Potter 1987), it has been suggested that these species would be more appropriately termed 'estuarine-opportunists' (Hedgpeth 1982; Lenanton and Potter 1987). A number of the marine teleosts which utilize estuaries as nurseries contribute to important commercial and recreational fisheries in either or both marine and estuarine waters. There is thus a clear need to preserve estuarine environments for the successful management of a number of important fisheries (e.g. McHugh 1976; Beal 1980). The importance of estuaries to commercial fisheries is emphasised by the fact that the 'estuarine-dependent' species of fish and crustaceans comprised 69% of the total weight landed by the commercial fishery of the United States in 1970 (McHugh 1967). Estuarine-dependent species have been estimated as contributing a similarly high proportion (70%) to the commercial fisheries in New South Wales, eastern Australia (Pollard 1976; 1981). Although this value for estuarine-dependent species is lower for south-western Australia and for Australia as a whole (20 and 32%, respectively), this still represents a large catch (Newell and Barber 1975; Lenanton and Potter 1987). In addition to commercial fishing, recreational fishing is an important activity in estuaries in many regions of the world (Caputi 1976; Lenanton 1979; Day *et al.* 1981; Marais 1988).

Factors influencing fish in estuaries

Several hypotheses have been invoked to explain why estuaries are used so extensively by the juveniles of marine fishes. For example, it has been suggested that, because estuaries are among the most productive environments in the world (Whittaker 1975; Correll 1978; Mann 1982), they supply an abundance of food, thereby facilitating the growth of juvenile fish. Furthermore, the presence of higher temperatures in estuaries than in marine waters would facilitate increased growth rates of fish in estuarine environments. In addition to the higher growth rates in estuaries, predation rates are believed to be lower in estuaries than the ocean. This view is supported by the fact that the incidence of large teleost piscivores is generally lower in estuaries than in marine waters (Blaber 1980; Blaber and Blaber 1980; Haedrich 1983). The macrophyte beds and turbid waters which are often found in estuaries are likely to increase protection from predation in these systems (Blaber and Blaber 1980; Lenanton *et al.* 1984; Cyrus and Blaber 1987a; b).

The number of species, density of individual species and the species composition of the fish community undergo seasonal changes in many estuaries (Gunter 1938; Dahlberg and Odum 1970; McErlean *et al.* 1973; Haedrich and Haedrich 1974; Livingston 1976; Quinn 1980; Bell *et al.* 1984; Claridge *et al.* 1986; Quinn and Kojis 1986). While these cycles have often been related to salinity and/or temperature, other variables, particularly in tropical or sub-tropical estuaries, may be more important in some cases. For example, in Moreton Bay in eastern Australia and the Lake St. Lucia estuarine system of South Africa, turbidity was important when spatial differences in salinity were relatively small (Blaber and Blaber 1980; Cyrus and Blaber 1987a; b). Similarly, in the absence of a strong salinity gradient, distance from estuary mouth had an important influence on the abundance and composition of juvenile fish associated with seagrass (*Zostera capricorni*) in the Hawkesbury River Estuary, eastern Australia (Bell *et al.* 1988). From the above, it can be seen that abiotic factors such as salinity, temperature, distance from estuary mouth and turbidity, and biotic variables such as aquatic vegetation, predation and competition, may affect the distribution and abundance of individual species and the structure of the fish communities of estuaries (see also Orth and Heck 1980; Young 1981; Orth *et al.* 1984; Pollard 1984).

Estuarine fish studies

Both the sampling methods used to investigate the fish fauna in estuaries and the range of estuarine environments sampled, have varied greatly among studies. For example, Thorman (1986) utilized a small beach seine to sample fish at sites along 60 km of the southern Bothnian Sea, whereas Little *et al.* (1988) used the same method of sampling at four sites in a tropical mangrove creek with no more than 5 km separating the most distant sites. In the former study, salinities were always less than 10‰, whereas they were very close to that of sea water in the mangrove creek (34-36‰). Differences in the scale of an investigation, such as those outlined above, can dramatically influence the results and hence the conclusions drawn from a study (Doherty and Williams 1988; Levin 1992). This point is often overlooked when comparisons are made between the findings from different studies.

In a review of estuarine fish studies, Haedrich (1983) made a plea for further work on the life history of fishes within estuaries, as well as highlighting the paucity of long term studies in these environments. Few studies of estuarine fish appear to have exceeded two years in duration (Haedrich 1983) and of those which have spanned a longer time period, most appear to have sampled only a limited number of sites or range of estuarine conditions. Thus, in a five year study of the fish fauna in a lagoon on the west coast of the United States, beach seines were used to catch fish at four sites, all with salinities very close to that of sea water, separated by distances of less than 2 km (Onuf and Quammen 1983). The six year investigation of Hillman *et al.* (1977) was also carried out at sites with salinities very similar to that of the ocean. Although significant seasonal variations in salinity were recorded during a five year study of the fish fauna in the Severn Estuary, United Kingdom, the major data source came from sampling at only one site (Claridge *et al.* 1986; Potter *et al.* 1986a). Very little is thus known of the relative importance of the influence of site within estuary, season and year on the abundance, and community structure of fish populations in estuaries.

Approaches to studies of fish populations in Western Australian estuaries

Work in the Swan Estuary of south-western Australia was undertaken over a five year period to gain a detailed understanding of the taxonomy and life histories of the fish found in this estuary; to investigate changes in the abundance of species and the composition of the fish fauna over the length of the system and with season and year; and to understand how different species were affected by changes in environmental variables, particularly salinity and temperature. These studies were jointly directed by Ian Potter of Murdoch University and Rod Lenanton of the Bernard Bowen Fisheries Research Institute, with funding from the Western Australian Departments of Fisheries, and Conservation and Environment.

The Swan Estuary covers a surface area of approximately 53 km^2 and is the second largest estuarine system in south-western Australia. This estuary and that of the Peel-Harvey (c 80 km south) are two permanently open systems on the west coast, whose mouths are separated by a distance of about 55 km. They are the most important estuaries for commercial and recreational fishing in this region (Lenanton 1979; Lenanton 1984; Lenanton et al. 1984).

The Swan Estuary comprises a long, narrow Entrance Channel that opens into extensive wide basins, which in turn lead into tidal, saline riverine areas, termed the lower, middle and upper estuary, respectively. Over the five year study, mean salinity ranged from 30‰ to 8.6‰ in the lower and upper estuary respectively (Loneragan et al. 1989). By contrast, mean temperature during this period was relatively constant (c 20°C) throughout the estuary. Variation in salinity increased with distance from the estuary mouth, whereas the variation in temperature was relatively stable in the different regions of the estuary. Both salinity and temperature were lower in the winter and spring than the summer and autumn. In the upper estuary, salinities can increase to about 30‰ during the late summer and early autumn, only slightly lower than salinities in the middle and lower estuary at this time. Other environmental variables such as turbidity, freshwater flow and nutrient levels are likely to vary in a similar way to the variation with distance from estuary mouth and season shown by salinity.

Large beach seines (swept area = 1 670 m^2 and 2 815 m^2) were used over a five year period to sample fish at ten sites in the shallow waters of the Swan Estuary (see Loneragan et al. 1989). Sampling for most of this study was at fortnightly or monthly intervals. In terms of distance from the estuary mouth, the sites were from 2 to 44 km upstream of the mouth.

The most abundant families of fish in the Swan were the Clupeidae, Terapontidae, Mugilidae, Apogonidae and Atherinidae. Of the 15 most abundant species in the shallows of the Swan Estuary, seven were marine teleosts which entered the estuary regularly and in large numbers (marine estuarine-opportunists), seven completed their life cycle within the estuary (estuarine) and one (*Nematalosa vlaminghi*) was anadromous. The contribution of individuals of the marine estuarine-opportunist category to catches in the shallows declined from nearly 95% in the lower estuary, to approximately 17% in the middle estuary and 6% in the upper estuary. The estuarine and anadromous groups together comprised 83 and 94% of the catches in the middle and upper estuaries, respectively.

The number of species and density of fish (measures of community structure) were influenced by distance from the estuary mouth, salinity and temperature; they declined with distance from estuary mouth and rose with increasing salinity and temperature. Classification and ordination distinguished the ichthyofauna of the saline reaches of the rivers from that of the lower reaches of the estuary. The faunal composition of the middle estuary of the Swan was also relatively distinct from those of the lower and upper estuary (Loneragan and Potter 1990). A secondary pattern of variation in the fish fauna, due to seasonal changes in composition, particularly in the upper estuary, was also detected using classification and ordination.

Site within the Swan Estuary generally influenced the densities of individual species to a greater extent than either Season or Year, or the interactions between these factors. When seasonal effects were important, they could be related to summer spawning migrations into the upper estuary (*Nematalosa vlaminghi*, *Amniataba caudavittatus*), spring recruitment of 0+ individuals into the lower estuary (*Mugil cephalus*) or winter movements into deeper and more saline waters (*Apogon rueppellii*). Marked annual variations in the density of *Torquigener pleurogramma* were related to large differences in the recruitment of the 0+ age class between years.

The studies within the Swan have helped to gain an understanding of the life histories and factors affecting fish populations over a broad spatial scale. More recent work has focussed on a finer spatial scale and has investigated the importance of macrophytes to fish in Wilson Inlet, a seasonally closed estuary on the south coast of Western Australia (Humphries *et al.* 1992).

Approaches to studies of fish populations in New South Wales estuaries

Much of the work in New South Wales has been carried out under the direction of Johann Bell and Dave Pollard of the NSW Fisheries Research Institute. Detailed studies of fish populations in estuaries have been completed in Botany Bay (e.g. Anon. 1981a; b) and in more recent years, concurrently in several estuaries along the coast of NSW (e.g. Ferrell and Bell 1991; McNeill *et al.* 1992; Worthington *et al.* 1992). Both these major studies were initiated in response to potential environmental impacts in coastal regions: extensions to Sydney airport in Botany Bay; and proposed development of port facilities in Jervis Bay for the Navy.

In general, the focus in these studies has been on establishing the importance of various habitats to fish and decapod populations, particularly seagrass compared with bare substrate. These studies have been carried out in regions of estuarine systems which appear to have a similar variation in salinity to that found in the lower Swan Estuary. The fish fauna is thus likely to be dominated by the juvenile stages of marine species. Estuarine and migratory (i.e. anadromous and catadromous) species are not likely to have been major contributors to the fish fauna in these studies.

The results of much of the work on the importance of seagrass systems to fish populations in Australia have been summarised by Bell and Pollard (1989). In their introduction they summarise the findings of previous studies and reviews with the following (modified from Bell and Pollard 1989):

- the diversity and density of fish are usually higher in seagrass than nearby bare areas;
- fish and decapods are found in seagrass for different lengths of time and at different stages of the life history;
- many fish species settle into seagrass from the plankton;
- seagrass, seagrass detritus and infauna of the seagrass are an underutilised food source for most species of fish compared to planktonic and epifaunal crustaceans;
- different species of fish are found in different positions in the seagrass canopy;
- the composition of the fish community and the relative abundance of different species can be influenced by the position of the bed in relation to other habitats and on the time of day; and
- the species composition of different seagrass beds often differs, even when the beds are adjacent.

The importance of different habitats to fish populations has been investigated by field sampling in single estuarine systems and more recently through concurrent sampling of a number of estuaries over the 18 months. The abundances of different species has then been correlated with attributes of the habitat, such as the shoot density of seagrass (e.g. Worthington *et al.* 1992). The results of these studies have been enhanced through the use of experimental field studies to test hypotheses about the importance of seagrass to fishes. This has involved studying settlement of larvae in artificial seagrass and examining the effect of manipulating the height and density of naturally occuring seagrass on the associated fauna (Bell et al. 1985; 1987; 1988; Bell and Westoby 1986).

From the general increase in abundance of fish and crustaceans with increased seagrass structure (i.e. higher shoot density, biomass) within a seagrass bed, it was proposed that predation and habitat selection were the main processes explaining these correlations (Heck and Orth 1980; Orth et al. 1984). More recently Bell and Westoby (1986) proposed an alternative hypothesis: that the distribution of fish and crustaceans among separate seagrass beds reflects the supply of larvae to those beds. They proposed that larvae settle to the first bed they encounter, regardless of the characteristics of the seagrass, and that individuals rarely leave the shelter of that bed during the following three to four months. Both the experimental field studies and the more descriptive work support the supply hypothesis. Worthington et al. (1992) conclude that the density of seagrass shoots explained little of the large scale variation in abundance of fish and decapods among separate seagrass beds and that their data support the supply model.

Summary and future directions

Work on the fish faunas found in estuaries is challenging due to the variability in both the types of species of fish found in these systems and in the environment itself. The approach of studying life histories has greatly improved our understanding of the dynamics of fish populations in estuaries of south-western Australia and the environmental factors which affect populations and the composition of the fauna. Over the relatively large spatial scales investigated in these studies, distance from estuary mouth and salinity (hence also the hydrology and characteristics of the catchment) have been identified as key factors affecting the composition of the fish fauna in the Swan Estuary.

Because of the detailed knowledge of fish populations in open estuaries and inshore waters of the west coast, it is timely to consider testing the generally held hypotheses about estuaries concerning their high productivity and refuge value for fish. Since the variation in the fish fauna is now well documented over a relatively large spatial scale, it is also appropriate to examine the importance of different habitat types in regions where environmental conditions are similar e.g. importance of seagrass in the lower and middle Swan system. Further work also needs to be undertaken to determine whether the factors affecting fish populations in open estuaries are the same in seasonally open and closed systems.

In the studies of fish populations of estuaries of New South Wales, most of the studies appear to have been undertaken at a smaller spatial scale within a particular system. At this level, variation in habitat type has an important influence on fish populations and community structure. Levin (1992) has suggested that the mechanisms affecting populations and communities vary according to the scale of the study and the hypotheses being investigated. This generality certainly seems to apply to studies of fish populations in estuaries on the east and west coasts of Australia. If we are interested in fish of the lower region of the estuary where salinity is relatively stable, then different habitat types and location of the habitats may determine the composition of the fish community and abundance of different species. However, if the hypothesis concerns the system as a whole, then distance from the estuary mouth and associated variables such as salinity are the key variables.

Bell and Pollard (1989) have discussed areas for new work on fish fauna in seagrass systems and suggested that the following areas were important:

- the effect of size, shape and location of the habitat within the estuary;
- the hypothesis that fish and decapods settle in the first bed they encounter and do not leave the bed; and

- differences in fish communities between different seagrass beds within the same estuary—are they due do the fact that the beds may be situated at different depths or to differences in the attributes of the seagrass? Does seagrass possess important attributes for fish popoulations or will different structures generate the same patterns in abundance?

In addition to these hypotheses, recent work by Underwood (1993) has challenged us to consider the design of studies on fish populations where the aim is to detect the effects of an impact. Can we apply his concept of multiple control estuaries when the catchment characteristics and morphology (hence hydrology and salinity regimes) of different systems vary greatly?

References

Anon. (1981a). The ecology of fish in Botany Bay—community structure. *State Pollution Control Commission of New South Wales.* Report No. *BBS23A.*

Anon. (1981b). The ecology of fish in Botany Bay—biology of commercially and recreationally important species. *State Pollution Control Commission of New South Wales.* Report No. *BBS23B.*

Beal, K.L. (1980). Territorial sea fisheries management and estuarine dependence. In: *Estuarine perspectives.* pp67-77. Kennedy, V.S. (Ed.) Academic Press, London.

Beckley, L.E. (1984). The ichthyofauna of the Sundays Estuary, South Africa, with particular reference to the juvenile marine component. *Estuaries* **7**, 248–258.

Bell, J.D. and D.A. Pollard (1989). Ecology of fish assemblages and fisheries associated with seagrasses. *In* Larkum, A.D. W., A.J. McComb and S. Shepherd (Eds) *A. Biology of seagrasses: A treatise on the biology of seagrasses with special reference to the Australian region,* pp565–609.

Bell, J.D. and M. Westoby (1986). Importance of local changes in leaf height and density to fish and decapods associated with seagrasses. *Journal of Experimental Marine Biology and Ecology* **104**, 249–274.

Bell, J.D., A.S. Steffe and M. Westoby (1985). Artificial seagrass: how useful is it for field experiments on fish and macroinvertebrates. *Journal of Experimental Marine Biology and Ecology* **90**, 171–177.

Bell, J.D., A.S. Steffe and M. Westoby (1988). Location of seagrass beds in estuaries: effects on associated fish and decapods. *Journal of Experimental Marine Biology and Ecology* **122**, 127–146.

Bell, J.D., M. Westoby and A. S. Steffe (1987). Fish larvae settling in seagrass: do they discriminate between beds of different leaf density? *Journal of Experimental Marine Biology and Ecology* **111**, 133–144.

Bell, J.D., D. A. Pollard, J. J. Burchmore, B.C. Pease and M.J. Middleton (1984). Structure of a fish community in a temperate tidal mangrove creek in Botany Bay, New South Wales. *Australian Journal of Marine and Freshwater Research* **35**, 33–46.

Blaber, S.J.M. (1980). Fish of the Trinity Inlet system of north Queensland with notes on the ecology of fish faunas of tropical Indo-Pacific estuaries. *Australian Journal of Marine and Freshwater Research* **31**, 137–146.

Blaber, S.J.M. (1985). The ecology of fishes of estuaries and lagoons of the Indo-Pacific with particular reference to southeast Africa. *In: Fish community ecology in estuaries and coastal lagoons: towards an ecosystem integration.* pp 247-266. Yáñez-Arancibia, A. (Ed.). Universidad Nacional Automoma de Mexico. ISBN 968-837-618-3.

Blaber, S.J.M. (1987). Factors affecting recruitment and survival of mugilids in estuaries and coastal waters of southeastern Africa. *Transactions of the American Fisheries Society Symposium* **1**, 507–518.

Blaber, S.J.M. and T.G. Blaber (1980). Factors affecting the distribution of juvenile estuarine and inshore fish. *Journal of Fish Biology* **17**, 143–162.

Caputi, N. (1976). Creel census of amateur line fishermen in the Blackwood River Estuary, Western Australia. *Australian Journal of Marine and Freshwater Research* **27**, 583–593.

Chrystal, P.J., I.C. Potter N. R. Loneragan and C.P. Holt (1985). Age structure, growth rates, movement patterns and feeding in an estuarine population of the cardinalfish *Apogon rueppellii. Marine Biology* **85**, 185–197.

Claridge, P.N., I.C. Potter and M.W. Hardisty (1986). Seasonal changes in movements, abundance, size composition and diversity of the fish fauna of the Severn Estuary. *Journal of the Marine Biological Association of the United Kingdom* **66**, 229–258.

Clarke, J., W.G. Smith, W. Kendall. and M.P. Fahay (1969). Studies of estuarine dependence of Atlantic coastal fishes. *United States Fisheries and Wildlife Services Technical Paper* No. **28**.

Cloem, J.E. and F.H. Nichols (1985). Time scales and mechanisms of estuarine variability, a synthesis from studies of San Francisco Bay. *Hydrobiologia* **129**, 229–237.

Correll, D.L. (1978). Estuarine productivity. *BioSciences* **28**, 646–650.

Cronin, L.E. and A. J. Mansueti (1971). The biology of the estuary. *In: A symposium on the biological significance of estuaries*, 14–39. Douglas, P.A. and Stroud, R.H. (Eds). Sport Fishing Institute, Washington, D.C.

Cyrus, D.P. and S.J.M. Blaber (1987a). The influence of turbidity on juvenile fishes in estuaries. Part 1. Field studies at Lake St. Lucia on the southeastern coast of South Africa. *Journal of Experimental Marine Biology and Ecology* **109**, 53–70.

Cyrus, D.P. and S.J.M. Blaber (1987b). The influence of turbidity on juvenile fishes in estuaries. Part 2. Laboratory studies, comparisons with field data and conclusions. *Journal of Experimental Marine Biology and Ecology* **109**, 71–91.

Dahlberg, M.D. and E.P. Odum (1970). Annual cycles of species occurrence, abundance, and diversity in Georgia estuarine fish populations. *American Midland Naturalist* **83**, 382–392.

Dando, P.R. (1984). Reproduction in estuarine fish. *In: Fish reproduction strategies and tactics.* pp 155–170. Potts, G. W. and R. J. Wootton (Eds.). Academic Press, London.

Davis, T.L.O. (1988). Temporal changes in the fish fauna entering a tidal swamp system in tropical Australia. *Environmental Biology of Fishes* **21**, 161–172.

Day, J.H. (1980). What is an estuary? *South African Journal of Science* **76**, 198.

Day, J.H. (1981). Estuarine currents, salinities and temperatures. *In: Estuarine ecology with particular reference to Southern Africa.* pp 27–44. Day, J.H. (Ed.). Balkema, Rotterdam.

Day, J.H., S.J. M Blaber and J. H. Wallace (1981). Estuarine fishes. *In: Estuarine ecology with particular reference to Southern Africa.* pp 197–221. Day, J.H. (Ed.). Balkema, Rotterdam.

Doherty, P.J. and D. McB Williams (1988). The replenishment of coral reef fish populations. *Oceanography and Marine Biology Annual Reviews* **26**, 487–551.

Ferrell, D.J. and J.D. Bell. (1991). Differences among assemblages of fish associated with *Zostera capricorni* and bare sand over a large spatial scale. *Marine Ecology Progress Series* **72**, 15–24.

Fourtier, L. and W.C. Leggett (1982). Fickian transport and the dispersal of fish larvae in estuaries. *Canadian Journal of Fisheries and Aquatic Sciences* **39**, 1150–1163.

Gunter, G. (1938). Seasonal variations in abundance of certain estuarine and marine fishes in Louisiana, with particular reference to life histories. *Ecological monographs* **8**, 314–346.

Haedrich, R.L. (1983). Estuarine fishes. *In: Ecosystems of the world* **26** : Estuaries and enclosed seas. pp 183-207. Ketchum, B.H. (Ed.). Elsevier Scientific Publishing Company, Oxford.

Haedrich, R.L. and S.O. Haedrich (1974). A seasonal survey of the fishes in the Mystic River, a polluted estuary in downtown Boston, Massachusetts. *Estuarine, Coastal and Marine Science* **2**, 59–73.

Heck, K.L. Jr. and R.J. Orth (1980). Seagrass habitats: the roles of habitat complexity, competition and predation in structuring associated fish and motile macroinvertebrate assemblages. *In: Estuarine Perspectives.* Kennedy, V. S. (Ed). pp 449–464. Academic Press, New York.

Hedgpeth, J.W. (1982). Estuarine dependence and colonisation. *Atlantica* **5**, 57–58.

Hillman, R.E., N. W. Davis and J. Wennemer (1977). Abundance, diversity and stability in shore-zone fish communities in an area of Long Island Sound affected by the thermal discharge of a nuclear power station. *Estuarine, Coastal and Marine Science* **5**, 355–381.

Humphries, P., I.C. Potter and N.R. Loneragan (1992). The fish community in the shallows of a temperate Australian estuary: relationships with the aquatic macrophyte *Ruppia megacarpa* and environmental variables. *Estuarine, Coastal and Shelf Science* **34**, 325–346.

Lenanton, R.C.J. (1977). Fishes from the hypersaline waters of the stromatolite zone of Shark Bay, Western Australia. *Copeia 1977*, 387–390.

Lenanton, R.C.J. (1979). The inshore-marine and estuarine licensed amateur fishery of Western Australia. *Fisheries Research Bulletin of Western Australia* **23**, 1–33.

Lenanton, R.C.J. (1982). Alternative non-estuarine nursery habitats for some commercially and recreationally important fish species of south-western Australia. *Australian Journal of Marine and Freshwater Sciences* **33**, 881–890.

Lenanton, R.C.J. (1984). The commercial fisheries of temperate Western Australian estuaries: early settlement to 1975. *Report of the Department of Fisheries and Wildlife of Western Australia* **62**, 1-82.

Lenanton, R.C.J. and I.C. Potter (1987). Contribution of estuaries to commercial fisheries in temperate Western Australia and the concept of estuarine dependence. *Estuaries* **10**, 28–35.

Lenanton, R.C.J., I.C. Potter, N. R. Loneragan and P.J. Chrystal (1984). Age structure and changes in abundance of three important teleosts in a eutrophic estuary. *Journal of Zoology, London* **203**, 311–327.

Levin, S.A. (1992). The problem of pattern and scale in ecology. *Ecology* **73**, 1943–1967.

Little, M.C., P.J. Reay and S.J. Grove. (1988). The fish community of an east African mangrove creek. *Journal of Fish Biology* **32**, 729–747.

Livingston, R.J. (1976). Diurnal and seasonal fluctuations of organisms in a North Florida estuary. *Estuarine, Coastal and Marine Science* **4**, 373–400.

Loneragan, N.R. and I.C. Potter (1990). Factors influencing fish community structure and distribution of different life-cycle categories in shallow waters of a large Australian estuary. *Marine Biology* **106**, 25–38.

Loneragan, N.R., I.C. Potter and R. C. J. Lenanton (1989). Influence of site, season and year on contributions made by marine, estuarine, diadromous and freshwater species to the fish fauna of a temperate Australian estuary. *Marine Biology* **103**, 461–479.

Mann, K.H. (1982). *Ecology of coastal waters*. University of California Press, Los Angeles, 322pp.

Marais, J.F.K. (1988). Some factors that influence fish abundance in South African estuaries. *South African Journal of Science* **6**, 67–77.

McComb, A.J., R.P. Atkins, P.B. Birch, D.G. Gordon and R.J. Lukatelich (1981). Eutrophication in the Peel-Harvey estuarine system. *In: Nutrient enrichment in estuaries*, pp 323–342. Nielson, B.J. and L.F Cronin (Eds). Humana Press, New Jersey.

McErlean, A.J., S.G. O'Connor, J.A. Mihursky and C.J. Gibson (1973). Abundance, diversity and seasonal patterns of esturine fish populations. *Estuarine, Coastal and Marine Science* **1**, 19–36

McHugh, J.L. (1967). Estuarine nekton. *In: Estuaries*. pp 581–620. Lauff, G.H. (Ed.). American Association for the Advancement of Science, Washington D.C.

McHugh, J.L. (1976). Estuarine fisheries: are they doomed? *In: Estuarine processes* **1**, 15–27. Wiley, M. (Ed.). Academic Press, New York.

McNeill, S.E., D.G. Worthington, D. J. Ferrell and J. D. Bell (1992). Consistently outstanding recruitment of five species of fish to a seagrass bed in Botany Bay, NSW. *Australian Journal of Ecology*, **17**, 359–365.

Newell, B.S. and W.E. Barber (1975). Estuaries important to Australian fisheries. *Australian Fisheries* **34**, 17–22.

Onuf, C.P. and M.L. Quammen (1983). Fishes in a California coastal lagoon: effects of major storms on distribution and abundance. *Marine Ecology Progress Series* **12**, 1–14.

Orth, R.J. and K.L. Heck Jr. (1980). Structural components of eelgrass, (*Zostera marina*) meadows in the lower Chesapeake Bay. Fishes. *Estuaries* **3**, 278–288.

Orth, R.J., K.L. Heck Jr. and J. van Montfrans (1984). Faunal communities in seagrass beds: a review of the influence of plant structure and prey characteristics on predator-prey relationships. *Estuaries* **7**, 339-350.

Pearcy, W.G. and S.W. Richards (1962). Distribution and ecology of fishes of the Mystic River Estuary, Connecticut. *Ecology* **43**, 248–259.

Pollard, D.A. (1976). Estuaries must be protected. *Australian Fisheries* **36**, 6-10.

Pollard, D.A. (1981). Estuaries are valuable contributors to fish production. *Australian Fisheries* **40**, 7-9.

Pollard, D.A. (1984). A review of ecological studies on seagrass-fish communities with particular reference to recent studies in Australia. *Aquatic Botany* **18**, 3-42.

Potter, I.C., P.N. Claridge and R.M. Warwick (1986a). Consistency of seasonal changes in an estuarine fish assemblage. *Marine Ecology Progress Series* **32**, 217-228.

Potter, I.C., W. Ivantsoff, R. Cameron and J. Minnard (1986b). Life cycles and distribution of atherinids in the marine and estuarine waters of southern Australia. *Hydrobiologia* **139**, 23–40.

Potter, I.C., L.E. Beckley, A.K. Whitfield and R.C.J. Lenanton (1990). Comparisons between the roles played by estuaries in the life cycles of fishes in temperate Western Australia and southern Africa. *Environmental Biology of Fishes* **28**, 143–178.

Prince, J.D., I.C. Potter, R.C.J. Lenanton and N.R. Loneragan (1982). Segregation and feeding of atherinid species (Teleostei) in south-western Australian estuaries. *Australian Journal of Marine and Freshwater Research* **33**, 865-880.

Pritchard, D. W. (1967). What is an estuary, physical viewpoint. *In: Estuaries*. pp 3-5. Lauff, G. (Ed.) American Association for the Advancement of Science, Washington, D.C.

Quinn, R.J. (1980). Analysis of temporal changes in fish assemblages in Serpentine Creek, Queensland. *Environmental Biology of Fishes* **5**, 117–133.

Quinn, R.J. and B.L. Kojis (1986). Annual variation in the nocturnal nekton assemblage of a tropical estuary. *Estuarine, Coastal and Shelf Science* **22**, 63–90.

Ross, S.W. and S.P. Epperly (1985). Utilization of shallow estuarine nursery areas by fishes in Pamlico Sound and adjacent tributaries, North Carolina. In: *Fish community ecology in estuaries and coastal lagoons: towards an ecosystem integration.* pp 207-232. Yáñez-Arancibia, A. (Ed.) Universidad Nacional Automoma de Mexico. ISBN 968-837-618-3.

Smale, M.J. (1984). Inshore small-mesh trawling survey of the Cape south coast. Part 3. The occurrence and feeding of *Argyrosomus hololepidotus*, *Pomatomus saltatrix* and *Merluccius capensis*. *South African Journal of Zoology* **19**, 170–179.

Thorman, S. (1986). Physical factors affecting the abundance and species richness of fishes in the shallow waters of the southern Bothnian Sea (Sweden). *Estuarine, Coastal and Shelf Science* **22**, 357-369.

Underwood, A.J. (1993). The mechanics of spatially replicated sampling programmes to detect environmental impacts in a variable world. *Australian Journal of Ecology* **18**, 99–116.

Whittaker, R.H. (1975). *Communities and ecosystems.* Macmillan, New York. 385pp.

Worthington, D.G., D.J. Ferrell, S. E. McNeill and J.D. Bell (1992). Effects of the shoot density of seagrass on fish and decapods: are correlations evident over larger spatial scales. *Marine Biology* **112**, 139–146.

Van den Broek, W.L.F. (1979). A seasonal survey of fish populations in the lower Medway Estuary, Kent, based on power station screen samples. *Estuarine, Coastal and Marine Science* **9**, 1–15.

Young, P.C. (1981). Temporal changes in the vagile epibenthic fauna of two seagrass meadows (*Zostera capricorni* and *Posidonia australis*). *Marine Ecology Progress Series* **5**, 91–102.

DISCUSSION OF SESSION 4

Recorded by M.I.Kangas
South Australian Department of Fisheries
GPO Box 1625
Adelaide SA 5001

Questions were addressed to individual speakers and followed by more general discussion.

After the presentation by *John Koehn*, Peter Jackson commented that when looking at the key variables determining fish populations in streams we have to accept that these key factors will differ between different river systems and that there is a need for 'regionalisation' of river systems. In addition, biotic interactions need to be considered. John Koehn agreed and highlighted the need to consider introduced species and their interactions.

Karen Edyvane pointed out that we should also be looking at community level indicators, like Karr's Index of Biological Integrity. John Koehn responded that to look at the integrity of the system you need to consider the whole system and all linkages, and this knowledge at all levels is lacking.

Jim Puckridge raised the point that one key factor that had been omitted in the preceding discussions was one of temporal variability, particularly hydrological variability in riverine environments. This factor needs to be recognised as it is one that is commonly under threat. Bryan Pierce agreed with Jim Puckridge's comments regarding freshwater and estuarine systems.

Murray Macdonald questioned whether *Peter Young's* definition of habitat in terms of physico-chemical attributes, without mentioning species interactions, was intentional or otherwise. Peter Young responded that he prefers to call the physico-chemical attributes an 'ecotome' rather that a habitat, the habitat actually being within an ecotome. His definition of a habitat is one of a fairly restricted area or 'patch' which can be defined by a number of various descriptors.

Russel Reichelt commented on the complexity of systems and the small amount of resources available to study the key factors and asked the panel to comment on some of the processes needed to filter and select priorities for research.

Neil Loneragan described the work that has been done in Western Australia regarding estuarine systems by defining levels of environmental variation in distinct regions of systems. The sort of questions they can be moving towards is looking at specific habitats within a zone. He noted that in estuarine work the linkages are poorly defined between estuaries and shallow inshore systems. In the case of the Peel Harvey system the future directions would come from some appropriate ecological modelling and making predictions and testing these predictions.

John Koehn emphasised that research priorities depend on what the question is. Managers come up with questions set at various levels, either on an ecosystem level or a specific question. For example the habitat work conducted in Victoria was essentially stimulated by funding of streamflow determination and from this came the recognition that we did not have the knowledge of the parameters required to plug into those sorts of models. It was also recognised that habitat data were required in order to manage things. In addition a lot of research, particularly in freshwater work is externally funded and these are specific projects; no-one will fund ecosystem projects.

Peter Young was of the view that experimental management is feasible for freshwater and estuarine systems. Modelling, such as that proposed in Port Philip Bay and done in Western Australia, will actually identify the critical things that we need to know. For offshore situations it is still guesswork. It is more likely that the effects of fishing on the habitat are more important compared to other things impacting on the habitat. We need to know what the gear is doing. John Koehn emphasised that in freshwater systems it is the changes to habitat or habitat degradation that are more of a problem than fishing.

Gina Newton focussed on an issue that had not been raised previously, linking estuarine and marine environments, that of the retention of larval and juvenile organisms, related to the geomorphology of the estuarine mouths in terms of the flow regimes. Also the stratification of those waters in estuaries.

Peter Jackson brought the discussion back to the question of the important variables to measure. He suggested that since there has been some excellent work in the freshwater systems by John Koehn and his team in Victoria and Angela Arthington in Queensland, an Australia-wide freshwater habitat group be established to see whether certain areas can be agreed on as needing work. Expanding on Gina Newton's reference to hydrological regimes he restated his earlier comments on classifying rivers according to their hydrological regime as being a starting point to putting groups of rivers together. Due to a lack of resources any work done should try to look at aspects that are as widely applicable as possible.

Bryan Pierce highlighted the fact that most of us are working in a 'reactive' mode so we cannot collect data that will allow us to 'shape' things in the future so that episodes such as that with the Murray cod do not occur again. He believed that we need to go beyond a reactive mode.

Peter Gehrke commented that there is a problem with defining a fish habitat but in all three systems we can group key factors as; limiting, controlling or masking. We should be able to come up with a suite of variables which are major controlling factors in one major habitat type and which factors are limiting. For some freshwater systems the limiting factors may change. He gave an example of this. He then invited the panellists to give a summary of what they consider limiting factors and controlling factors for their various systems.

John Koehn responded that there certainly are critical and limiting factors in freshwater systems. We can group rivers which have similar attributes such as upland versus lowland. To determine critical factors we need to consider each species; some factors may involve several species but some factors may affect only one species. It is not easy to come up with just a few factors. On further thought he stated that there are some critical stages that could be concentrated on, such as the critical factors affecting spawning inducement and larval feeding.

Neil Loneragan said that for the estuarine systems, looking at the habitat level, seagrass and algal species are limited by flow and light regime. Key factors for fish species become more complex depending on whether you are considering the community level, recreational species, commercial species etc. Factors influencing fish communities are salinity regime, distance from estuary mouth, rainfall and water flow, but he was not sure whether these can be classified as key factors.

Peter Young responded that for estuarine systems, salinity would definitely be a factor. In the open ocean, zoogeography would play a part as would depth. There is not enough knowledge to state key factors there.

John Glaister tried to focus the discussion towards considering when fisheries managers are making decisions about fish habitats - what do they need to know? Jenny Burchmore supported John Glaister's comments and expanded as a manager she is confident in the current information available that seagrasses are important as fish habitats and that more emphasis needs to be placed on other less studied areas such as the impact of dredging.

Roland Pitcher raised the need for modelling to help highlight priorities in research and asked Norm Hall to explain modelling in more detail. Norm Hall responded that models try to put together people's understanding of processes in a very simple form. Modelling plays a part in trying to identify the questions you need to ask and the data that are needed, and then trying to put them together and finding areas where the model doesn't actually work. In other words modelling is the formality of putting your thoughts down on paper, testing them and trying to improve on your knowledge.

Bryan Pierce then tried, as chairperson, to bring the threads of the discussion together. We have tried to elucidate the key factors that link fish and their environment. For all the three environments discussed we know they are complex and that the detailed knowledge of key factors for many species is limited. Any model produced would be preliminary. The only place where habitat is not a major management issue in terms of human impact is in the offshore environment. Everywhere else it is a critical issue and needs to be continually focussed on, whether we do it through modelling and taking the predictions and testing them, or through adaptive management in the field. We need to come up with results that mean more fish in the water because in all habitats we are looking at a continuing decline trend.

Jeremy Prince thought that marine fisheries scientists can learn from the freshwater scientist in such areas as methods of describing fish habitats. He stressed the need to understand the spatial and temporal structure of fish stocks to lay the framework for measuring and monitoring abundance in the field instead of just monitoring the CPUE (catch per unit effort).

Campbell Davies noted that for coral reef communities there is very strong evidence of habitat preferences. When attempting stock assessment, spatial heterogeneity must be considered.

GENERAL DISCUSSION—DAY 1

Recorded by D.C. Smith
Victorian Fisheries Research Institute
PO Box 114
Queenscliff VIC 3225

Stan Moberly began the discussion with the comment that although we don't know as much about the environment as we would like, we probably don't appreciate how important our knowledge of its complexities is. He stressed the importance of a proactive approach to getting the information across to the relevant people, rather than simply reacting through, for example, the Environmental Impact Statement (EIS) process. He commented that not very much discussion has taken place on how we get the information across to those who impact upon the environment.

Chairperson Bryan Pierce called for comments on the key issue of communication. Barbara Richardson argued that there was an initial step before disseminating information: the need for a structure of where we are going with research and its application. There needs to be some statement on what are now thought to be the critical driving mechanisms in fish habitat. This, together with statements by managers on what they consider the major threats, could form a structure to enable priority setting. It would also allow assessment year by year on progress.

Stan Moberly thought the time was right now and wondered whether the Society was going to take a lead approach in dealing with this issue. Barbara Richardson responded by saying that she was not proposing delaying this, but hoped that an outcome of this workshop would be a "state of our understanding". This might provide a tool for getting the message across but it would also act as a guiding framework for looking at how we are going in a year, five years or ten years time. She agreed with Stan Moberly that it is extremely important to get the message out there, to create the awareness and profile, and to get the involvement and participation of the wider community.

Gary Jackson agreed with both speakers. He said that Phil Cadwallader had written what was wrong with native fish populations 20 years ago, yet nothing had really eventuated, except that more populations were under threat and more habitat had been lost. He was concerned that in ten years time we should not be in the same situation. Information must be in a form that can be used by fisheries managers and also to educate the public.

Rob Lewis, speaking from a manager's perspective, mentioned his reverse chair technique. When, for example, industry complains or asks for action on a particular issue, he puts them in his chair and moves to the other side of the table and asks them what they would do. He suggested that this might be applicable here. He discussed the need to find compatibility and agreement on our preferred approach, i.e. getting our own house in order first. He posed the question as to how he, as a fisheries manager, can start to influence the decision makers.

In the light of earlier comments about the Society's role, John Glaister proposed a small group to meet at the end of the day to develop a

Society position that could be discussed on the final day. Barbara Richardson agreed to chair the group and several others were "volunteered".

John Koehn agreed with earlier speakers arguing that it is important that we are involved in the decision-making process, because if we are not, the best information will not be utilised. His second point was that we always assume that people in other fields know more than us. From his experience that just isn't true.

John Koehn went on to describe his own involvement with foresters and the positive outcomes that such interactions can give. Initially they were uncooperative, and didn't want to hear that, for example, sedimentation was a problem even though they realised it was, but eventually they asked for advice on how to stop the sediments hitting the stream. It was a daunting task, because there really aren't the 3 star scientifically replicated 10 year experiments to give exact answers. But from the knowledge and principles already available, the important points are there. We came up with all sorts of things and now there's even a suggestion that we will get another code of forest practice revised. The foresters have now come to accept that we are the water experts, they are the tree experts and so we are in there. It has taken up a huge amount of time but you can have an effect and you can change the thinking. However, it is a long hard education process. We have gone some way to getting aquatic fauna, as opposed to just water, included into forestry studies.

Peter Young reiterated that it comes down to "how to influence the decision makers?". He mentioned that the Society has a high profile with the fisheries decision makers and is recognised, for example, by the Standing Committee on Fisheries and Aquaculture. The Society is consulted on many issues but could probably be further consulted if it chose to be. Where we have a low profile, however, is with other groups. He posed the questions: How many University courses in forestry or agriculture include the impact of their activities on the aquatic environment? Have we considered approaching Vice Chancellors and suggesting that they might perhaps change some of their courses? To what extent can the Directors of Fisheries influence, say, the Directors of Forestry? He commented that increasing our profile in this area could be very difficult particularly given Australia's federal system. Another approach the Society hasn't really considered is through the media. It can be a risky business but it certainly is going directly to the people. He mentioned that CSIRO are commencing training for selected staff in dealing with the media, recognising that this is the way of the future. Bryan Pierce supported the need to "sell" our fisheries better.

Dianne Hughes argued that communication between Societies is important. She also suggested that an outcome of the workshop could be a report or review that could be disseminated widely to all groups, including engineers, foresters, government and parliamentary committees, Universities etc. She stressed the importance of the Society setting down its goals and objectives and stating how these might be achieved. This point was also taken up by Bob O'Boyle who supported the need for the Society to put out a "state of the habitat" report. This was a positive approach, considering that all he had heard so far were comments about how bad things were. He suggested that a group could start working towards it over the next year.

Stan Moberly discussed the American Fisheries Society's experience with the Forestry Service and the benefits of an integrated approach. Members of the Society together with fisheries people in the Forestry Service and some foresters convinced the Service that current practices had a considerable impact on the aquatic environment. This has seen, over the last 6-7 years, the fish and wildlife budget of the Forestry Service increase about four times, and their staff of fisheries and wildlife personnel also increase dramatically. Although not all the answers are known, this was a start in the right direction. Trees will still be cut down, but if they can be cut in a way that has less impact on aquatic resources, that is progress.

John Koehn mentioned Victorian reports which included state of rivers. He remarked that representing fish has an advantage. People, including foresters, are interested in fish, they fish for them, eat them etc. But foresters know about trees and we know about fish, but conversely we don't know about trees and they don't know about fish. He stressed the need to bridge this gap in knowledge. He also suggested that there was a general interest in popular stories in the media, mentioning Stuart Rowland on the ABC National program "Australia All Over" talking about cod, and the videos put out by the Murray-Darling Basin Commission. Information on key species can also be put in a form that managers can use easily. People are receptive to this. Fish scientists have to get involved with all groups, especially in freshwater where you have foresters, engineers and water managers.

Mick Olsen endorsed John Koehn's multi-disciplinary approach but cautioned that fisheries managers often do not like to see other disciplines coming into their area of responsibility, yet everybody is aiming at sustaining the habitat and wildlife. A further advantage of having a multi-disciplinary thrust at maintaining habitat is that you may be able to get extra funds.

Duncan Leadbitter talked about the role of the fishing industry in habitat protection. He argued that 10 years ago fishermen were castigated for not being involved in habitat and now they were zealots. He put forward a word of caution, however, in terms of education, on the top down approach that science is the fountain of all knowledge. He argued that he had wasted 6 months when starting his job of trying to convince fishermen of the importance of habitat because they already knew this. What they needed to know was how to approach Government. Fishermen need training in how the planning system works, how environmental impact assessment works, how they can have their say and how they can get habitats protected. So different sectors will need different information. For example, it will be technical information for the foresters, and information on the planning process for commercial fishermen (and presumably the same for recreational fishers).

Peter Gehrke's view of the day's proceedings was that everybody believes that habitat is important but there are no definitions on how to delineate where we are going. If our goals are not well delineated, how are we going to identify the problems we are going to encounter in achieving those goals? Scientists have vested interests in learning more and more details about the systems they are addressing and, from professional and numerous other perspectives, that is the correct approach. However, from the point of view of managing the habitats that support the fishery, we have got to be prepared to take a more reductionist view to the systems we are looking at. We will never understand all the variables but at least in the conceptual models that we use we can summarise the state of knowledge. Even flow charts of some sort would be of use in providing guide-lines for what we have to do. We have to define the key processes driving the habitats that we are looking at. In many of the fields we are dealing with, and from a modelling perspective, answers to within an order of magnitude are adequate. As many of our processes operate on a logarithmic or exponential scale, to try and reach 5% or 10% resolution is not practicable. We know a lot more and we can give a lot more advice than we have been doing to date. In summary, a more reductionist view is needed in the way we approach research for management as we get down to defining what is the important scale of sustaining habitats for specific fisheries issues.

Bryan Pierce concluded the discussion by agreeing that we have got to set some goals and that in the past we haven't. This is, perhaps, why we aren't talking within the same framework all the time. Key factors - clearly we do not have them - and a reductionist view would certainly get us there as long as they are critically tested. Picking up Stan Moberly's point, he thought that probably the most critical factor that wasn't dealt with in the organisation of today was the issue of communicating whatever we decide, to the public, to make something actually happen. This should be dealt with on the second day of the workshop.

SESSION 5

IMPACT OF HUMAN ACTIVITIES ON HABITAT AND FISHERIES

Session Chairperson: C.M. Macdonald
Session Panellists: M. Mallen-Cooper
G.P. Jenkins
R.A. Campbell
Rapporteur: P.C. Coutin

CHAIRPERSON'S INTRODUCTION

C. M. MacDonald

Marine Science Laboratories
P.O. Box 114
Queenscliff VIC 3225

Good morning and welcome to Session 5 on Day 2 of this workshop. During yesterday afternoon's sessions we had extensive discussions on the relationship between aquatic organisms and their habitats, and on the possible impacts of various types of human activities on these relationships. This morning I hope to further develop the theme of the workshop by broadening the discussion to consider linkages not only between various human activities and habitat alteration, but also between such habitat alterations and fluctuations in the abundance or availability of economically important species.

The main objectives of the discussion will be:

- to identify major categories of human-induced impact on aquatic habitats;

- to assess, for each of these major categories, the quantity and quality of scientific evidence for linkages between human activities, habitat alterations and fluctuations in fishery resources; and

- to identify (if possible) strategies for demonstrating these linkages in order to convince governments and the general public of the importance of protecting and maintaining aquatic habitats.

I will begin by presenting a list of human-induced impacts on aquatic habitats which are considered to be of some significance in the Australian context (Table 1). This list can be added to or modified during the course of the discussion.

Before introducing the panellists for this session, I would like to share with you a brief impression of the impact of human activities on coastal marine and estuarine habitats and fisheries in the USA as a prelude to discussion of the Australian situation. This impression is based on information collected over more than 20 years by the US National Marine Fisheries Service (NMFS) as part of its National Habitat Conservation Program. I was fortunate enough to attend an international coastal management symposium (Coastal Zone '89) in the US at which this information was presented, and I subsequently obtained a transcript of the presentation and permission to cite the information, courtesy of the Office of Protected Resources and Habitat Programs, NMFS, Washington D.C.

The overwhelming impression from the NMFS presentation was that as population growth - and therefore increased urban, industrial and agricultural activities - has occurred along the US coastline over the last 100 years or more, so have there been increased problems in maintaining the health and productivity of inshore and coastal aquatic habitats and living resources. Over 110 million acres, or more than 50%, of *total* US wetland habitats had been lost up to the mid-1970's. The total loss of *coastal* wetlands is not as severe (about 20% up to the mid-1970s), but the rate of loss has accelerated to more than 100,000 acres per year in recent years. This is of particular concern when about 70% of all US commercial fishery resources are estuarine-dependant, and estuarine-dependant species support commercial and recreational fisheries each estimated to generate economic benefits of several billion dollars per year.

The NMFS presentation identifies physical habitat modification, contaminant loading, excessive nutrient loading, water diversions/obstructions, waste disposal, pathogen introductions, chemical loading, marine debris, dredging and spoil disposal, hydropower and forestry and mining as the most important sources of impact on US coastal and inshore marine habitats. Specific evidence of the impact of human activities on aquatic habitats and/or fish populations included:

- decline of Columbia River Basin salmon and steelhead trout catches to 16-25% of historic levels following a loss of access by wild fish to 33% of historic headwater spawning areas due to hydropower dams;

- 60-80% decline in striped bass populations in San Francisco Bay following increasing diversion of freshwater inflows and loss of up to 80% of Bay wetlands;

- 80% or more decline of commercial catches of striped bass, river herring and American shad in Chesapeake Bay during 1960-1983. This coincides with substantial seagrass loss associated with excessive nutrient loading, and with other human-induced habitat modifications in brackish water nursery areas;

- a substantial proportion of US coastal waters under shellfish aquaculture in the early 1960s have since been closed - mainly because of contamination from human sewage effluent. Many of these closed areas have high levels of human sewage tracers in the sediments; and

- US coastal areas with the highest human population densities are also the areas where the highest levels of contaminants (eg. DDT, PCBs, PAHs) are being found in fish and sediments. These are also the areas where fish have the highest incidence of cancerous growths and other serious diseases.

The NMFS presentation concludes by stressing that while fish populations can recover from overfishing, the pollution and/or loss of fish habitats results in long-term and generally irreversible population losses.

This sobering impression of the situation in the USA may not be immediately relevant to us because of Australia's much smaller population and economy. However, Australia's population continues to grow and, like the USA, we have an affluent life style and a strong coastal orientation. The US trends outlined above can therefore be viewed as an indicator of what is in store for Australian aquatic habitats and living resources if the population and economic trends of recent decades continue.

So what do we know of the links between human activities, changes in aquatic habitats and production of fishery resources in Australia? My experience in drawing together panellists for this session suggests that our 'knowledge' in this area consists of a great deal of hypothesis, some circumstantial evidence, and only a small amount of rigorous quantitative assessment of such links. Perhaps some of you in this workshop might have a different perception, in which case we would like to hear it.

We will now prime this discussion session by hearing three Australian case studies of habitat change and fluctuations in fishery resources. Martin Mallen-Cooper will talk about habitat changes and fluctuations in the distribution and abundance of freshwater fish, mainly in eastern Australian drainages. Greg Jenkins will then discuss the relationship between seagrass loss and declines in commercial scalefish catches in Westernport Bay, Victoria. Robert Campbell will then conclude by describing the effects of trawling on marine benthic habitats and fish communities on the North West Shelf of Australia. It is hoped that these presentations will stimulate general discussion on the relationship between human activities, habitat changes and fisheries production, which will in turn lead to the formulation of findings and/or conclusions regarding the discussion objectives listed above.

Table 1. Major categories of human impact on Australian aquatic habitats

Freshwater:

Physical habitat loss (flow regulation, development, channelisation etc.)

Excess nutrients/eutrophication

Erosion/sediment deposition

Salinisation (land use practices)

Contaminants (organic and inorganic)

Species introductions/translocations

Marine:

Physical impact of fishing methods

Hydrocarbons (oil)

Waste dumping/debris

Sewage/industrial effluent disposal

Estuaries/Bays:

Physical habitat loss (shoreline dredging and spoil disposal, etc.)

Excess nutrients/eutrophication

Catchment erosion/sediment deposition

Species introductions/translocations

Physical impact of fishing methods

Contaminants (including heavy metals and hydrocarbons)

Global:

Greenhouse effect/ climate change

Ozone depletion/UV damage

HABITAT CHANGES AND DECLINES OF FRESHWATER FISH IN AUSTRALIA: WHAT IS THE EVIDENCE AND DO WE NEED MORE?

M. Mallen-Cooper

Fisheries Research Institute
P.O. Box 21
Cronulla NSW 2230

Introduction

Many freshwater fish species in Australian river systems have declined in their range and abundance during the last 100 years and habitat changes are frequently considered to be the major cause. In this paper I am going to briefly examine the published evidence for these declines, the habitat changes which are implicated, and the quality of the evidence linking the two.

While there is general consensus among fish biologists that many native freshwater fishes now have a reduced distribution and abundance compared with pre-European settlement, the published evidence is sparse. Records of commercial catch have been used to document the decline of Murray cod (*Macullochella peeli*) in New South Wales (Rowland 1989), barramundi (*Lates calcarifer*) in Queensland (Pollard *et al.* 1980), and silver perch (*Bidyanus bidyanus*) in the Murray-Darling river system (T.J. Johnson pers. comm.). In Victoria, museum records, published reports and interviews with anglers and biologists have been used to document the decline of Murray cod, trout cod (*M. macquariensis*), Macquarie perch (*M. australasica*) and golden perch (*M. ambigua*) (Cadwallader 1981; Cadwallader and Gooley 1984; Brumley 1987).

Habitat loss and degradation by human activites have been hypothesized by many authors as the main cause of declines in the abundance and distribution of freshwater fish in Australian rivers (e.g. Lake 1971; Cadwallader 1978; Pollard *et al.* 1980; Merrick and Schmida 1984; Koehn and O'Connor 1990; Lloyd *et al.* 1991). However, what is the real evidence linking the two?

In Table 1 have been listed the major elements of habitat that have been affected by human activities with an estimate of the quality of the evidence linking these with declines of native fish in Australia.

Dams and weirs as barriers to fish movement

The first element of habitat I have listed is 'access' (to suitable habitat), which is a basic need of freshwater fish. In modified river systems, movement of fish along streams and onto floodplains is restricted to varying degrees by dams, weirs, levee banks, road crossings and culverts. In the streams of coastal south-eastern Australia half of the aquatic habitat has been obstructed by man-made barriers (Harris 1984). The disappearance of many migratory species above large dams is readily acknowledged, but there are few published studies documenting these declines in Australian rivers, probably because the evidence linking the adverse impacts of the impoundment on fish migration is clear-cut. A notable example is the disappearance of natural populations of golden perch from the upper River Murray following the construction of Hume Dam and Yarrawonga Weir (Lake 1971). Low weirs can be submerged by floodwaters which provide some passage for fish. However, these barriers still very seriously restrict fish movement. One of the few published studies demonstrating this is by Harris (1988) who sampled with gill nets above and below a tidal weir (Liverpool) near Sydney and found no migratory species above the weir and abundant fish below the weir. A less quantitative but notable example of this type of impact is the decline of the commercial catch of barramundi in Queensland attributed to tidal barrages preventing migration (Pollard et al. 1980). Despite the lack of published information, the restriction of fish movements particularly at large dams and weirs is obvious and the links between the human activity, habitat change and declines of fish are good.

River regulation

River regulation has had direct and measurable effects on the flow and temperature regimes of streams in Australia (Walker 1985; Cadwallader 1986). However the causal links between these changes and declines in freshwater fish populations are difficult to demonstrate. The evidence is usually anecdotal and difficult to separate from other changes in habitat. Some evidence is provided by Rowland (1989) who linked the decline of Murray cod in NSW in the 1950's and 60's to the reduction in flooding caused by major dams in the tablelands. Similarly, Harris (1988) demonstrated that flooding was important in the reproduction of Australian bass (*Macquaria novemaculeata*) and linked the decline in recruitment of this species with flood suppression caused by dams.

Other experimental research on the biology of native fish has produced stronger, but still indirect, evidence linking river regulation and declines of fish. Specifically, flooding and temperature have been identified in fish hatcheries as an important cue for spawning of native species (Lake 1967; Rowland 1983). The effects of river regulation can be direct and obvious, such as a fish kill below a dam when the flow is stopped completely and fish are stranded (Bishop and Bell 1978). Fish kills have also been reported when floodgates are opened on coastal streams releasing poor quality water (Richardson 1981). However, the direct evidence for river regulation causing a decline in freshwater fish in Australia is generally absent.

Records of fish movement through fishways provide some evidence of declines in fish movement and abundance due to changes in flows caused by river regulation. The numbers of golden perch using the Euston (Lock 15) fishway in the River Murray have declined by 43% over the last 50 years, while the movement of silver perch has declined by 93% (Figure 1). Over the same period the small floods (5,000-10,000 ML/day) that stimulate migration in these fish have declined by approximately half (Close 1990). In this case, river regulation appears to have reduced the stimulus to migrate, although other factors such as water quality may also have contributed. For silver perch, and perhaps for golden perch also, the reduced numbers provide some quantitative evidence of the decline in their abundance. These declines cannot be attributed solely to river regulation, as this is very difficult to separate from other habitat changes such as siltation and barriers to migration.

Changes in water quality

In some cases changes in water quality through human activities have produced visible and direct evidence such as fish kills from chemical pollutants. Toxicological studies in Australia have also produced indirect evidence from laboratory studies of the effects of pollutants on fish (e.g. Baker and Walden 1984; Gehrke 1988). Sewage and fertilisers have decreased water quality through nutrient enrichment and increased algal growth, leading to reduced oxygen in the water (Williams 1980). In Australia, this has probably changed the composition of fish communities and reduced the distribution of some native species, but there appears to be no published evidence.

Fish kills from acidic water have been described overseas (e.g. Leivestad and Muniz 1976). In Australia acidic soils and runoff from inappropriate land use have been implicated in fish kills but it has only been reported in the media and not in the scientific literature. Acidic water has also been implicated in red-spot disease in fish but again there is no published evidence in the scientific literature. Although there is some indirect evidence from laboratory work for the need for high water quality (e.g. Gehrke and Fielder 1988) the evidence linking changes in water quality with declines of freshwater fish in Australia is generally poor.

Substrate

Soil erosion resulting from land clearance, agriculture, timber harvesting and other activities has led to increased turbidity and suspended solids. When these suspended solids settle, the substrate of the stream is modified by siltation and sedimentation. There is strong direct and indirect evidence from overseas research that such habitat changes severely affect fish, particularly the early life stages (Alabaster and Lloyd 1980; Campbell and Doeg 1989).

However, in Australia there are only two reported studies of the effect of turbidity and sediment on fish; Richardson (1985) found a decrease in *Galaxias maculatus* populations following forestry operations in southern New South Wales, and Koehn *et al.* (1991) reported that in artificial conditions, eggs of freshwater blackfish *Gadopsis marmoratus*, spotted galaxias *Galaxias truttaceus*, climbing galaxias *G. brevipinnis* and Macquarie perch *Macquaria australasica* showed high mortalities when smothered with a fine layer of silt. River channelization directly affects the substrate. Although there is much evidence from research overseas that channelization is detrimental to fish (e.g. review by Swales 1982) the evidence for adverse effects on native fish in Australian rivers is limited to one study by Hortle and Lake (1982). These researchers compared fish of channelized and unchannelized sections of the Bunyip River in Victoria and found that fish numbers, biomass and species richness were all reduced in the channelized sections. Channelization, however, is not widespread in Australia compared with catchment erosion and siltation.

Instream cover and riparian vegetation

Instream cover and riparian vegetation have frequently been removed to improve navigation or channel capacity, or for sand and gravel extraction. This important element of habitat can be replaced by introduced aquatic plants. There are only two studies examining the relationship between freshwater fish and instream cover in Australia; Hortle and Lake (1983), discussed above, and Arthington *et al.* (1983). The latter study related the loss of native aquatic plants and the increase in introduced plants in streams near Brisbane to a decline in the distribution of five native fish species.

Apart from these two studies the direct evidence linking declines in instream cover and riparian vegetation with declines in fish abundance in Australian streams is poorly documented, although again there is reasonable evidence from studies overseas (e.g. Angermeir and Karr 1984; Fausch and Northcote 1992). There is, however, indirect evidence through life history studies of Australian fishes which describe the use of habitat by fish and hence the inherent value of the habitat before it is removed or degraded (e.g. Pollard 1971; Cadwallader and Rogan 1977; Harris 1988).

Conclusions

In other countries there has been considerable research carried out and published which links human activities, changes in habitat and declines of freshwater fish. In Australia there has been little such research published. It may be possible to apply the broad principles from the overseas studies to the management of Australian freshwater ecosystems, but we still need to carry out and publish research which quantifies the responses of fish species and populations to the environmental conditions typical of coastal and inland rivers in Australia. To answer the fundamental questions asked by managers, such as how much of a particular impact is acceptable in a particular habitat, there is an urgent need for research in the following areas:

- determination of habitat utilisation by fish and derivation of key habitat requirements, such as instream cover, substrates, role of aquatic and riparian vegetation, and use of floodplains;

- flow needs of fish - for breeding, movement, dispersal and recruitment;

- experimental evidence is needed on the levels of silt and sediment tolerated by fish at all life stages, and their responses to key water quality variables; and

- fish passage requirements - fishways have been developed which are suitable for some species (Mallen-Cooper 1992), but the requirements of many species are unconfirmed or unknown. In addition, a broader knowledge of the swimming ability of native fishes is needed to design access onto floodplains and through culverts and pipes in road crossings.

To maintain fish populations which are viable in the long-term, and to prevent the further decline of populations of threatened species, we need to devise new approaches to the management of aquatic and riverine ecosystems in Australia. As part of this process, we need to understand the full environmental consequences of activities which are contributing (or have contributed in the past) to the current decline in fish diversity and abundance. The research priorities listed above are essential to this understanding.

Acknowledgements

I thank Jenny Burchmore, Drs John Harris, Jane Mallen-Cooper and Stephen Swales for comments on the manuscript.

References

Alabaster J.S. and R. Lloyd (1980). *Water quality criteria for freshwater fish.* (FAO: Butterworths, London).

Angermeier, P.L. and J.R. Karr (1984). Relationships between woody debris and fish habitat in a small warmwater stream. *Transactions of the American Fisheries Society* **113**, 716–726.

Arthington, A.H., D.A. Milton and R.J. McKay (1983). Effects of urban development and habitat alterations on the distribution and abundance of native and freshwater fish in the Brisbane region, Queensland. *Australian Journal of Ecology* **8**, 87–101.

Baker, L. and D. Walden (1984). Acute toxicity of copper and zinc to three fish species from the Alligator Rivers region. *Technical Memorandum of the Supervising Scientist for the Alligator Rivers Region* No. 8, 27 pp.

Bishop, K.A. and J.D. Bell (1978). Observations on the fish fauna below Tallowa Dam (Shoalhaven River, New South Wales) during river flow stoppages. *Australian Journal of Marine and Freshwater Research* **29**, 543–549.

Brumley, A.R. (1987). Past and present distributions of golden perch *Macquaria ambigua* (Pisces: Percichthyidae) in Victoria, with reference to releases of hatchery-produced fry. *Proceedings of the Royal Society of Victoria* **99**, 111–116.

Cadwallader, P.L. (1978). Some causes of the decline in range and abundance of native fishes in the Murray-Darling River system. *Proceedings of the Royal Society of Victoria* **90**, 211–224.

Cadwallader, P.L. (1981). Past and present distributions and translocations of Macquarie perch *Macquaria australasica* (Pisces: Percichthyidae), with particular reference to Victoria. *Proceedings of the Royal Society of Victoria* **93**, 23–30.

Cadwallader, P.L. (1986). Flow regulation in the Murray River system and its effects on the native fish fauna. In I.C. Campbell (Ed.) *Stream protection: the management of rivers for instream uses*. [Melbourne]: Water Studies Centre, Chisholm Institute of Technology.

Cadwallader, P.L. and G.J. Gooley (1984). Past and present distributions and translocations of Murray cod *Maccullochella peeli* and trout cod *M. macquariensis* (Pisces: Percichthyidae) in Victoria. *Proceedings of the Royal Society of Victoria* **96** (1), 33–43.

Cadwallader, P.L. and P.L. Rogan (1977). The Macquarie perch, *Macquaria australasica* (Pisces: Percichthyidae), of Lake Eildon, Victoria. *Australian Journal of Ecology* **2**, 409–418.

Campbell, I.C. and T.J. Doeg (1989). The impact of timber harvesting and production on streams: a review. *Australian Journal Marine and Freshwater Research* **40**, 519–539.

Close, A. (1990). The impact of man on the natural flow regime. In: *The Murray* (Murray-Darling Basin Commission: Inprint Ltd, Brisbane Qld.) pp. 61–74.

Fausch, K.D. and T.G. Northcote (1992). Large woody debris and salmonid habitat in a small coastal British Columbia stream. *Canadian Journal of Fisheries and Aquatic Science* **49**, 682–693.

Gehrke, P.C. (1988). Acute cardio-respiratory responses of spangled perch, *Leiopotherapon unicolor* (Günther 1859), to sublethal concentrations of zinc, temephos and 2,4-D. *Australian Journal of Marine and Freshwater Research* **39**, 767–774.

Gehrke, P.C. and D.R. Fielder (1988). Effects of temperature and dissolved oxygen on heart rate, ventilation rate and oxygen consumption of spangled perch, *Leiopotherapon unicolor* (Gunther 1859), (Percoidei, Teraponidae). *Journal of Comparative Physiology B* **157**, 771–782

Harris, J.H. (1984). Impoundment of coastal drainages of south-eastern Australia, and a review of its relevance to fish migrations. *Australian Zoologist* **21**, 235–250.

Harris, J.H. (1988). The demography of Australian bass, *Macquaria novemaculeata* (Perciformes, Percichthyidae) in the Sydney Basin. *Australian Journal of Marine and Freshwater Research* **39**, 355–369.

Hortle K.G. and P.S. Lake (1983). Fish of channelized and unchannelized sections of the Bunyip River, Victoria. *Australian Journal of Marine and Freshwater Research* **34**, 441–450.

Koehn, J.D. and W.G. O'Connor (1990). *Biological information for management of freshwater fish in Victoria*. (Government Printer, Melbourne) 165 pp.

Koehn, J., B. O'Connor and D. O'Mahoney (1991). The effects of sediment on fish. Proceedings of the 1991 annual conference of the Australian Society for Limnology. Held at Lorne, Victoria. (Abstract).

Lake, J.S. (1967). Rearing experiments with five species of Australian freshwater fishes. I. Inducement to spawning. *Australian Journal of Marine and Freshwater Research* **18**, 155–173.

Lake, J.S. (1971). *Freshwater fishes of Australia*. Nelson (Australia), Sydney. 61 pp.

Leivestad, H. and I.P. Muniz (1976). Fish kill at low Ph in a Norwegian river. *Nature (Lond.)* **259**, 65–72.

Lloyd, L., J. Puckridge and K. Walker (1991). The significance of fish populations in the Murray-Darling system and their requirements for survival. In: *Conservation in management of the River Murray system – making conservation count*. Proceedings of the Third Fenner Conference on the Environment, Canberra, September 1989, 86–99.

Mallen-Cooper, M. (1992). Swimming ability of juvenile Australian bass, *Macquaria novemaculeata* (Steindachner), and juvenile barramundi, *Lates calcarifer* (Bloch), in an experimental vertical-slot fishway. *Australian Journal of Marine and Freshwater Research* **43**, 823–834.

Merrick J.R. and G.E. Schmida (1984). *Australian freshwater Fishes* (Griffin press: Netley, South Australia).

Pollard, D.A. (1971). The biology of the landlocked form of the normally catadromous salmoniform fish *Galaxias maculatus* (Jenyns), I) Life cycle and origin. *Australian Journal of Marine and Freshwater Research* **22**, 91–123.

Pollard, D.A., L.C. Llewellyn and R.D.J. Tilzey (1980). Management of freshwater fish and fisheries. In: *An ecological basis for water resource management*. 227-270. (ANU press, Canberra, Australia).

Richardson, B.A. (1981). Fish kill in the Belmore River, Macleay River drainage, NSW, and the possible influence of flood mitigation works. In 'Proceedings of the Floodplain Management Conference, Canberra, ACT, Australia, 7-10 May, 1980'. Vol. 3 background papers. Australian Water Resources Council conference series, No. 4. (Australian Government Publishing Service: Canberra).

Richardson, B.A. (1985). The impact of forest road construction on the benthic invertebrate and fish fauna of a coastal stream in southern New South Wales. *Australian Society for Limnology Bulletin* **10**, 65–88.

Rowland S.J. (1983). Spawning of the Australian freshwater fish Murray cod, *Maccullochella peeli*, in earthen ponds. *Journal of Fish Biology* **23**, 525–534.

Rowland S.J. (1989). Aspects of the history and fishery of the Murray cod, *Maccullochella peeli* (Mitchell) (Percichthyidae). *Proceedings of the Linnean Society of N.S.W.* **111**, 201–213.

Sanger, A.C. (1990). Life history of the two-spined blackfish, *Gadopsis bispinosus,* in King Parrot Creek, Victoria. *Proceedings Royal Society of Victoria* **102**, 89–96.

Swales, S. (1982). Environmental effects of river channel works used in land drainage improvement. *Journal of Environmental Management* **14**, 103–126.

Walker, K.F. (1985). A review of the ecological effects of river regulation in Australia. *Hydrobiologia* **125**, 111–129.

Williams, W.D. (1980). Water as a waste transport and treatment mechanism: an ecological evaluation. *In: An ecological basis for water resource management.* 155–160. (ANU press, Canberra, Australia).

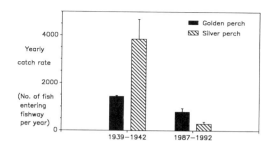

Figure 1. Yearly catch rate (mean ± s.e.) of golden perch and silver perch from the Euston fishway on the River Murray over periods 1939–1942 and 1987–1992.

Table 1. Impacts of human activities on fish habitat and the quality of the evidence linking these impacts with declines in fish abundance and distribution in Australia

Habitat	Human activity	Quality of evidence for causing decline in native fish distribution or abundance
access (to habitat)	dams, weirs, levee banks	good
flow and temperature regimes	river regulation	indirect
water quality	domestic, industry, agriculture	some good, some poor
substrate	siltation, sedimentation, channelization	poor
instream cover and riparian vegetation	de-snagging, increasing channel capacity	poor

ECOLOGICAL BASIS FOR PARALLEL DECLINES IN SEAGRASS HABITAT AND CATCHES OF COMMERCIAL FISH IN WESTERN PORT BAY, VICTORIA

G.P. Jenkins, G.J. Edgar, H.M.A. May and C. Shaw

Victorian Institute of Marine Sciences and Department of Zoology
University of Melbourne
P.O. Box 138
Queenscliff VIC 3225

A decline of over 70 % of the seagrass cover in Western Port Bay occurred between 1973 and 1984 (Bulthius et al. 1984; Shepherd et al. 1989). The losses occurred predominantly from intertidal areas of *Heterozostera tasmanica* (Bulthius et al. 1984; Shepherd et al. 1989). Although the exact causes of this decline are not known, the major proximate cause was thought to be desiccation and/or high temperatures coupled with a fine coating of adherent mud on the leaves reducing light levels (Bulthius et al. 1984; Shepherd et al. 1989). The underlying causes of losses may have included increased turbidity and sediment deposition resulting from catchment erosion and dredging operations, increased emersion at low tide due to changes in topography and tidal hydrology, and unusually high temperatures over summer (Bulthius 1983; Bulthius et al. 1984). Losses may have been self-perpetuating; initial seagrass death may have led to mudbank erosion, increasing suspended solids and sediment deposition in adjacent areas.

The decline in seagrass cover was paralleled by a decline of about 40 % in total commercial fish catches from Western Port Bay (Figure 1; MacDonald in press). Catches in Port Phillip Bay, excluding pilchards for which catches increased rapidly, remained relatively constant over the same period. Catch declines in Western Port Bay were particularly apparent for some species, such as leatherjackets (Figure 2) and grass whiting (Figure 3) where catches in the 1980's were at historically low levels (MacDonald in press). Although alternative explanations for the declines, such as overfishing, and changes in fishing effort or larval input, cannot be rejected, the results are suggestive of a link with seagrass decline for some species. In contrast, catches of other species such as yellow-eye mullet either showed no signs of decline, or actually increased (Figure 4). King George whiting showed a more complex pattern, with a major peak in catches in the early 1970's declining to approximately pre-peak levels in the 1980's (Figure 5). A similar early 1970's peak was apparent for King George whiting catches in Port Phillip Bay, suggesting that factors such as changes in effort, or larval input, were responsible for the peak (C.M. MacDonald, unpublished). Post-peak catches of King George whiting in Port Phillip Bay were approximately double those of pre-peak levels, while in Western Port Bay, pre- and post-peak levels were similar, suggesting a relatively greater decline in Western Port Bay.

This presentation describes a small part of a major research program on juvenile and adult fishes associated with seagrass beds in Port Phillip Bay, Western Port Bay and Corner Inlet aimed at identifying possible linkages between fish populations and seagrass habitats. For the purposes of this presentation we will concentrate on the distribution, abundance and diets of the four species described above to investigate possible reasons for the patterns of decline or otherwise of commercial catches in parallel with seagrass loss.

Results are described for sampling conducted in 1989/1990 at three sites in Swan Bay and one site on the adjacent coast of Port Phillip Bay (Figure 6). Seagrass and unvegetated habitats were sampled at each site. Sites at Queenscliff and St Leonards were adjacent to the shoreline whilst sites at Tin can and North Jetty were in deeper water (1 m MLWS). Dietary studies were conducted on fish collected at three subtidal sites in Western Port Bay; Rhyll, French Island and Tooradin.

Field sampling was conducted with fine-mesh seine nets of 10 and 15 m length and 1 mm mesh. The nets were small enough to selectively-sample specific habitats. When nets were deployed from a small boat, ropes were hauled using detachable weights to stop the net from rising from the bottom until completely retrieved.

Small juveniles of six-spined leatherjacket (Figure 7) and grass whiting (Figure 8) were almost exclusively collected from subtidal *Heterozostera* beds. Both species recruited directly to seagrass in the spring/summer (Figures 7 and 8). King George whiting recruited to unvegetated patches amongst subtidal *Heterozostera* in Swan Bay in spring, and older juveniles were collected in reasonable numbers from the near-shore Queenscliff site in February/March (Figure 9). Yellow-eye mullet were collected at the near-shore St Leonards and Queenscliff sites and were never collected at the deeper subtidal sites (Figure 10). This species tended to occur over unvegetated sand at St Leonards, but there was no obvious habitat preference at Queenscliff.

The diets of six-spined leatherjackets were dominated by seagrass associated biota, including both plant and animal material (Figure 11A).

The diets of grass whiting were dominated by seagrass-associated epifauna, mainly molluscs and crustaceans (Figure 11B). The smallest King George whiting juveniles consumed crustacean plankton which was quickly replaced by crustacean epifauna (Figure 12A), consisting mainly of groups such as epibenthic harpacticoid copepods, mysids and tethygeneid amphipods, which would tend to be concentrated in unvegetated patches amongst seagrass beds. Larger individuals consumed soft-sediment crustacean and polychaete infauna (Figure 12A). The diets of small juveniles of yellow-eye mullet were dominated by planktonic crustacea; however, the diets of larger individuals were composed mainly of algae and seagrass-associated epifauna (Figure 12B).

Six-spined leatherjackets and grass whiting are obviously highly dependent on seagrass in terms of habitat and diet. These species were not found to settle in areas where seagrass has been lost and replaced by unvegetated habitats. Food available in unvegetated habitats is probably also unsuitable for these species although this hypothesis would be difficult to test. In general, the parallel decline between seagrass and populations of these species is not surprising.

The link between seagrass and King George whiting is more subtle. Although post-larvae of this species mainly recruit to and feed in unvegetated patches, it is likely that seagrass detritus would lead to elevated abundances of prey species, possibly leading to elevated growth and survival of juvenile whiting. Swan Bay, with its greater amounts of macrophyte detritus, has already been shown to support higher meiofaunal abundances, and feeding rates of juvenile flounder, on unvegetated habitats compared with Port Phillip Bay (Shaw and Jenkins 1992). The infauna consumed by older individuals may also be more abundant in unvegetated sediments enriched by macrophyte detritus. Although decreased prey abundances due to loss of seagrass may be linked to the greater post-peak decline in catches of King

George whiting in Western Port Bay relative to Port Phillip Bay, in general, catches of this species may also be strongly influenced by other factors. In particular, recruitment of King George whiting is dependent on larval input from Bass Strait, and could be greatly influenced by interannual variability in Bass Strait current patterns (Jenkins and May, unpublished).

Juvenile yellow-eye mullet were not dependent on seagrass for habitat or diet, apparently foraging in shallow, near-shore areas for crustacean zooplankton, irrespective of benthic habitat. Although older individuals consumed some seagrass-associated epifauna, this linkage was apparently not critical to feeding success of this group.

To summarise, the strong parallel decline in fish catches and seagrass loss occurred in species which were specifically adapted to life in a seagrass habitat. Species with a reduced ecological link with seagrass habitat did not show a clear parallel decline with seagrass. The post-peak decline in King George whiting catches was greater in Western Port Bay than Port Phillip Bay, suggesting that the link between seagrass habitat and juvenile feeding may be of importance, although the pattern was dominated by other factors.

Work such as that described above identifies key factors which may vary with species, areas and times. However, there may also be useful broad patterns which may help determine which habitats should be protected for fisheries production purposes. For example, on a bay-wide scale, seagrass beds in Port Phillip Bay of similar structure vary in levels of recruitment depending on hydrodynamics (Jenkins, unpublished). This situation is similar to that described in New South Wales estuaries (Bell and Westoby 1986; Bell *et al.* 1987). It may be possible to predict which areas will receive high recruitment using hydrodynamic modelling (Jenkins and Black, unpublished).

Determining priority in preservation of habitats such as seagrass beds will depend on the particular values of interest. For example, in Port Phillip Bay, the seagrass habitats which would be preserved to maximise recruitment of economically-important fish species would be different to those preserved to maximise the overall diversity of fish species.

An important question is how to convey information such as that presented here to coastal managers and others who may influence possible impact on important habitats. Scientists are often reticent to become involved with the media; their views are often presented less than faithfully. However, other user or interest groups which are perceived by the public to conduct 'research' have a strong voice in the media. Although the sentiments expressed by such groups may be laudable in terms of habitat protection, unfortunately the information presented is often highly erroneous. A case in point was recent comments on the proposed relocation of a chemical storage facility from Coode Island, near Melbourne, to Point Wilson, near Geelong, on Port Phillip Bay. Claims were made by a lobby group on prime-time television news that most of the seagrass beds in Port Phillip Bay were located at Point Wilson and that these beds formed the basis of the food chain for dolphins in the bay. These claims were patently false and endangered the reputation and standing of scientists in general. Scientists could certainly do no worse than these groups. Perhaps professional scientific organisations should employ scientists trained specifically in media relations to convey correct information to the public.

Acknowledgements

We thank Murray MacDonald, Victorian Department of Conservation and Natural Resources, for comments on the manuscript and for providing data on fish catches in Western Port Bay. Funding for this research was provided by the Fishing Industry Research and Development Council.

References

Bell, J.D. and M. Westoby (1986). Variation in seagrass height and density over a wide spatial scale: effects on common fish and decapods. *J. Exp. Mar. Biol. Ecol.* **104**, 275–295.

Bell, J.D., M. Westoby and A.S. Steffe (1987). Fish larvae settling in seagrass: do they discriminate between beds of different leaf density? *J. Exp. Mar. Biol. Ecol.* **111**, 133–144.

Bulthius, D.A. (1983). A report on the status of seagrasses in Western Port in May 1983. Marine Science Laboratories, Queenscliff, Victoria. Internal Report No. 38, 5 pp.

Bulthuis, D.A., D.M. Axelrad, A.J. Bremner, N. Coleman, N.J. Holmes, C.T. Krebs, J.W. Marchant and M.J. Mickelson (1984). Loss of seagrasses in Western Port. Progress Report No. 1, December 1983 to March 1984. Marine Science Laboratories, Queenscliff, Victoria. Internal Report No. 73, 10 pp.

MacDonald, C.M. (1991). Fluctuations in seagrass habitats and commercial fish catches in Westernport Bay and the Gippsland Lakes, Victoria. *In* Recruitment Processes, Australian Society for Fish Biology Workshop, Hobart, 21 August 1991, D.A. Hancock (ed.), Bureau of Rural Resources Proceedings No. 16, AGPS, Canberra.

Shaw, M., and G.P. Jenkins (1992). Spatial variation in feeding, prey distribution, and food limitation of juvenile flounder, *Rhombosolea tapirina* Günther. *J. Exp. Mar. Biol. Ecol.* **165**, 1–21.

Shepherd, S.A., A.J. McComb, D.A. Bulthius, V. Neverauskas, D.A. Steffensen and R. West (1989). Decline of Seagrasses. *In* A.W.D. Larkum, A.J. McComb and S.A. Shepherd, *Biology of Seagrasses*. Elsevier, Amsterdam.

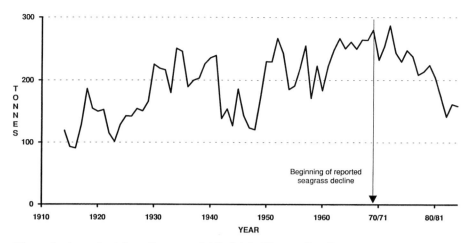

Figure 1. Annual catches of commercial finfish in Western Port Bay.

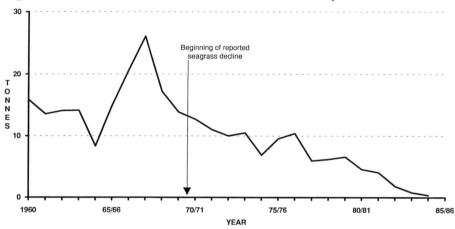

Figure 2. Annual catches of leatherjackets in Western Port Bay.

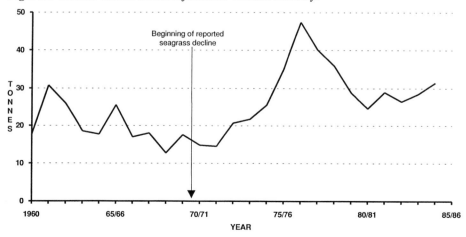

Figure 3. Annual catches of grass whiting in Western Port Bay.

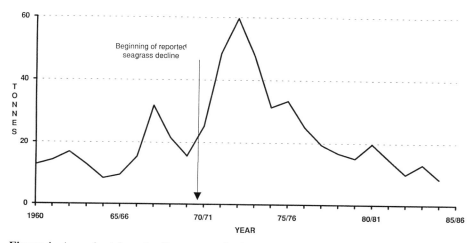

Figure 4. Annual catches of yellow-eye mullet in Western Port Bay.

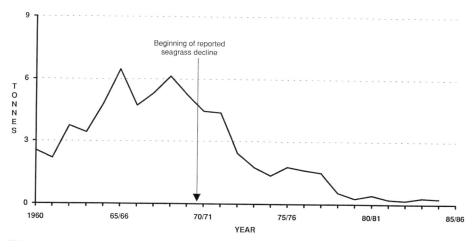

Figure 5. Annual catches of King George whiting in Western Port Bay.

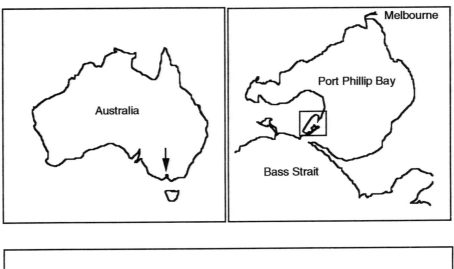
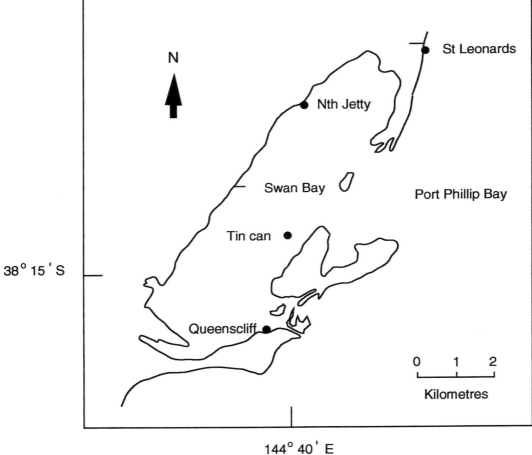

Figure 6. Location of sampling sites in Swan Bay and Port Phillip Bay, Victoria, from which juvenile fish were collected. Insets: Location of Port Phillip Bay on the Australian coast, and location of Swan Bay in Port Phillip Bay.

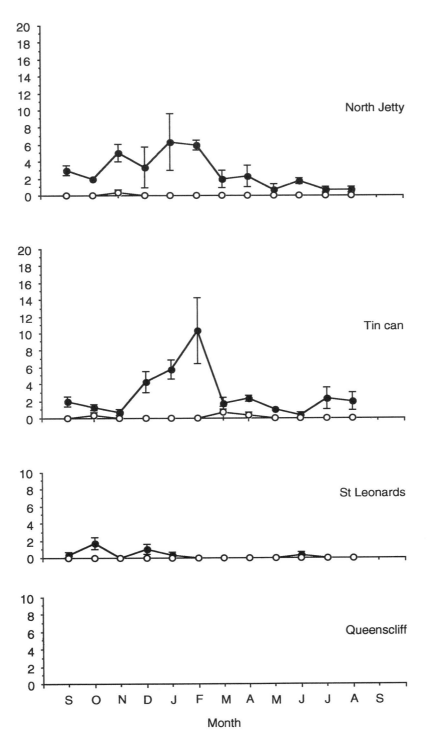

Figure 7. Abundance of juvenile six-spined leatherjackets in seine-net samples. Open circles, unvegetated habitat; closed circles, seagrass habitat.

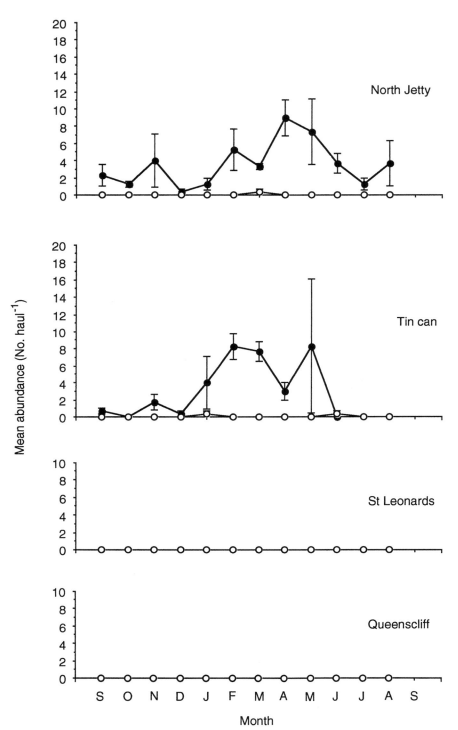

Figure 8. Abundance of juvenile grass-whiting in seine-net samples. Open circles, unvegetated habitat; closed circles, seagrass habitat.

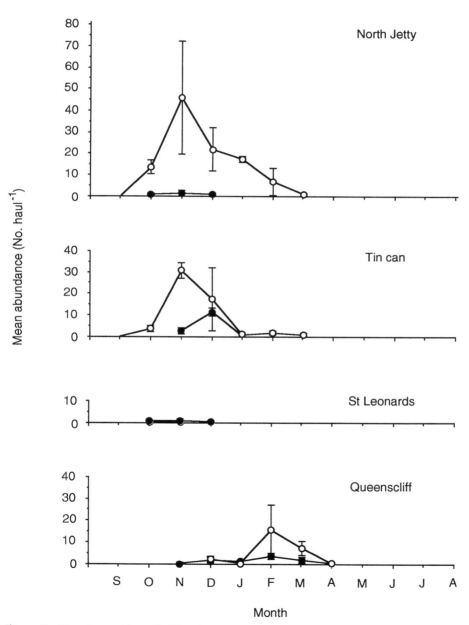

Figure 9. Abundance of juvenile King George whiting in seine-net samples. Open circles, unvegetated habitat; closed circles, seagrass habitat.

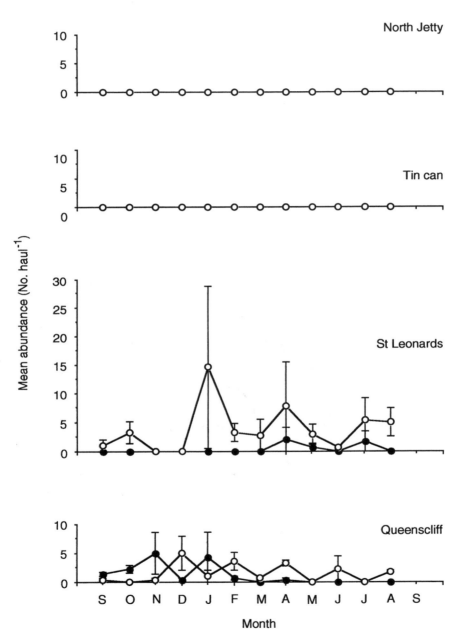

Figure 10. Abundance of juvenile yellow-eye mullet. Open circles, unvegetated habitat; closed circles, seagrass habitat.

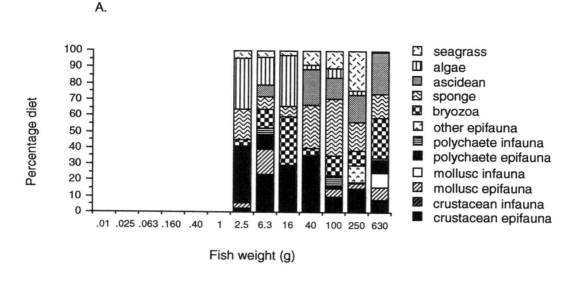
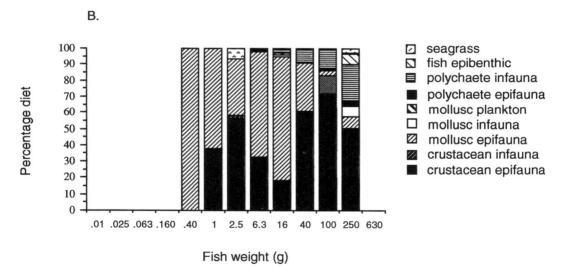

Figure 11. Dietary composition of fishes from Western Port Bay. A. Six-spined leatherjacket, B. Grass whiting.

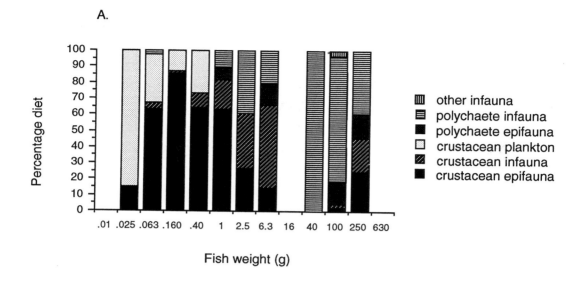
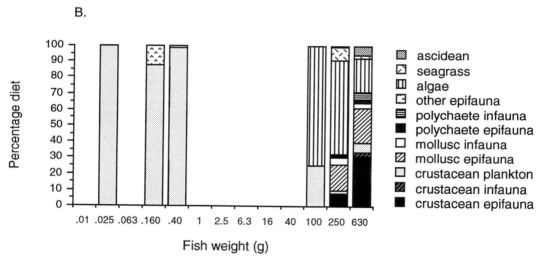

Figure 12. Dietary composition of fishes from Western Port Bay. A. King George whiting, B. Yellow-eye mullet.

EFFECTS OF TRAWLING ON THE MARINE HABITAT ON THE NORTH WEST SHELF OF AUSTRALIA AND IMPLICATIONS FOR SUSTAINABLE FISHERIES MANAGEMENT

K.J. Sainsbury, R.A. Campbell and A.W. Whitelaw
CSIRO Division of Fisheries
GPO Box 1538
Hobart TAS 7001

Introduction

Fish resources may be used in a variety of ways. The policy challenge for management is to allocate these resources between competing uses when members of the community with rights of access wish to use them in incompatible ways. Making a management decision will involve compromise between optimising the community benefits from various users of these resources and minimising the detrimental effects resulting from each type of use, so as to maximise the chances of long-term sustainable use of the resources. For most fish resources, however, there is considerable uncertainty about both the dynamics of the resource and the effects on it of exploitation, both direct or indirect. Furthermore, there is the need to identify the linkages between the fish resources and the habitats which support and nourish them.

Threats to fish resources in the past have focused on the problems of overfishing, with management responses leading to constraints being placed on fishing operations, most notably through strategies to limit harvest rates. More recently however, there has been concern about the destruction of the habitats which are required to support many of our fish resources. Generally, this concern has focused on the effects of fishing itself on the marine habitat.

An early reference to this problem appeared in 'Guide to the Fishmarket', written by Dr. Bellamy, a Cornishman, as long ago as 1842.

"Fishing, taken generally, interferes in the slightest way with the habits of the creatures in question; but the employment of a trawl, during a long series of years, must assuredly act with the greatest prejudice towards them. Dragged along with force over considerable areas of marine bottom, it tears away, promiscuously, hosts of inferior beings there resident, besides bringing destruction on the multitudes of smaller fishes, the whole of which, be it observed, are the appointed diet of those edible species sought as human food. It also disturbs and drags forth the masses of deposited ova of various species. An interference with the economical arrangement of creation, of such magnitude and of such duration, will hereafter bring its fruits in a perceptible diminution of these articles of consumption for which we have so great necessity. The trawl is fast bringing ruin on numbers of poorer orders requiring the most considerable attention."

Today, sustainable utilisation of our marine resources has become a high priority. While the concerns about sustainability may not be new, the need to deal with them effectively has perhaps never been greater because of increasing demands placed on marine resources. Although management decisions are made in the knowledge of the 'best available scientific advice', individual decisions are often still demonstrated to be incorrect by the excessive depletion of some fish resources.

In this paper some of the issues concerning sustainable use of the marine fish resources through sustaining their habitat are illustrated. The paper is in three parts. The first identifies some of the problems facing managers of Australia's tropical marine fish resources, the second reviews the North West Shelf fishery and describes some of the changes to the ecological communities in this region, and the third describes both the research and the management approach taken in this region. While much of this has been described elsewhere (see Sainsbury 1987; 1988; 1991), the focus here will be on both the effects of trawling on the marine habitat and the relationship between this habitat and the fish species composition found on the North West Shelf.

Tropical multispecies fisheries

1. The management problem

It is widely recognized that fishing activities modify the structure and dynamics of ecological communities. For example, changes in the relative abundance of species in exploited communities are common (Hempel 1978; Cushing 1980). Furthermore, the decrease of a few species may be critical if these are the only ones catchable or marketable. On the other hand, widespread changes in community composition may be irrelevant if most species are catchable or have similar market value. Unfortunately the changes that happen to communities which include exploited species are not reliably predictable, and consequently the economic and social effects of such changes are also not reliably predictable. In most situations the dynamics of marine communities are poorly understood and reasons for the observed changes in an exploited community are hard to determine. Typically, ecological theory can be used to derive a number of different hypotheses about the mechanisms and dynamics of change that are all consistent with the observations of an exploited community. However, there is considerable uncertainty about the mechanisms which should be emphasized in a community model and there are also limitations imposed by the information available. Consequently, prediction of a marine community's response to changes in fishing activities resulting from management decisions can remain highly uncertain.

While much attention in the past has focused on the fisheries in temperate Australia, within the last decade the tropical fisheries of northern Australia have been receiving more attention. Unlike the fisheries in temperate waters, Australian tropical fisheries are normally characterised by a large number of species (generally several hundreds) in which commercial interest is usually shown in only a small portion. In addition to the monitoring difficulties caused by this high species diversity, the biology of most species is largely unknown and there are numerous interactions between the species themselves. While predator-prey relationships and competition for food resources will be important in determining the response of the community to exploitation, the actual role these interactions play in the dynamics of the community can, at most, only be guessed at. Finally, the fishing mortality is usually not equal for all species in the community and can be highly influenced by the targeting behaviour of the fishers. For example, some species are specific to a particular depth range and if this depth range is highly targeted by the fishing effort, the fishing mortality on these species can be high while it remains low on the more widespread species (Sainsbury 1982).

These features of tropical multispecies fisheries combine to increase the uncertainties in scientific advice to managers. A common approach to management under uncertainty is the 'certainty equivalent' approach in which a 'best model' of the resource is selected as the most likely of the available possible models and used to derive management actions that are then applied to the resource as a whole. Alternatively, consideration can be given to a number of alternative ecological interpretations of the available data, and the consequences of the possible management options can be evaluated across all these alternative models. With the 'certainty equivalent' approach the resource is managed as if the most likely available model is true; it ignores the uncertainties in understanding about resource dynamics and usually results in the whole resource being treated under the same management regime. This can increase the risk to the resource if an error is made, and future observations of the resource usually have little power to discriminate between different hypotheses about resource dynamics because possible control variables are confounded. Another approach is based on the methodology developed by Walters and Hilborn (1976; 1978) and Walters (1986) for management of resources with poorly known dynamics. This approach, often termed 'actively adaptive management', attempts to develop a 'reasoned empiricism' by managing the resource in a way which permits accelerated identification of the response of the resource to exploitation and provides guidance to the selection of long-term management options. In the following we discuss the application of this actively adaptive management approach to a tropical fish community off northwestern Australia.

2. The North West Shelf—A case study

The character of Australia's tropical multispecies fisheries and the related management problems are well illustrated by the fishery located on Australia's North West Shelf region (Figure 1). The area supports a diverse and productive demersal fish community and has been fished mostly by foreign distant-water fleets (Sainsbury 1987).

Major commercial fishing first took place on the North West Shelf between 1959 and 1963 when Japanese stern trawlers targeted the large fish stocks of the genus *Lethrinus*. During the next three years, over seven thousand tonnes of this genus were caught with a catch rate around 500 kg/hour. Species of this single genus comprised about half of the catch. In 1972 Taiwanese pair trawlers began operations in the area. The fishing was initially very intensive with annual catches of between 20,000 and 30,000 tonnes for the first five years. The retained catch mostly comprised the genera *Nemipterus*, *Saurida*, *Lutjanus* and *Lethrinus*. In 1979 the area came within the Australian Fishing Zone and consequently under Australian management. While the Taiwanese continued to trawl under a licence arrangement with the Australian Government, a small domestic trap fishery began operation in the western part of the shelf in the early 1980's mainly in areas subjected to little trawling (Moran *et al.* 1988).

The Taiwanese continued to fish the North West Shelf until 1989, by which time a small domestic trawl operation had also become established. However, in this time several changes to the biotic community on the Shelf had taken place. First, a major change within the species composition of the fish catch was observed, although the total catch rate of the trawl fishery remained relatively constant. During the initial years the larger species groups, dominated by the genera *Lutjanus* and *Lethrinus*, accounted for 40% to 60% of the catch, while two other genera, *Nemipterus* and *Saurida*, together comprised around 10%. By the mid-1980's, however, this situation had reversed with *Lutjanus* and *Lethrinus* species comprising about 10% of the catch, and *Nemipterus* and *Saurida* around 25%. Secondly, the demersal habitat of the Shelf was also known to have altered during this period. The quantity of epibenthic fauna (mostly sponges, alcyonarians, and gorgonians) caught in trawls was observed to be considerably less than that recorded prior to and during the early development of the pair trawl fishery. Based on research data, the average catch rate of sponges fell from around 500 kg/hour to only a few kilograms/hour (Sainsbury 1987).

Changes in the relative species composition of catches on the North West Shelf was of little consequence to the Taiwanese pair trawl fishery, as all retained species had a similar commercial value (Liu and Lai 1980). However, the domestic Australian fishing operations have no market for *Nemipterus* and *Saurida*, and rely heavily on *Lethrinus* and *Lutjanus* species for commercial viability. With the introduction of Australian management after 1979, options for maximising the involvement of the domestic industry were considered, including trap fishing. Expansion of the trap fishery, however, seemed possible only if there was a return to something like the historical fish community composition. Furthermore, the poor state of knowledge about the resource dynamics made prediction of the resource's response to major alterations in the historical management regime highly uncertain. It was not clear whether or not the fish community could recover its earlier composition or what management actions would be best in attempting a recovery.

3a. CSIRO Study—1982–1983

A central management issue arising from the fish community changes observed on the North West Shelf (NWS) was the fact that while the species composition had initially been attractive commercially, the present day composition was not. Given this situation, three questions arose in response to the needs of management:

1) What caused these changes in species composition?

2) Is it worth trying to reverse the change?

3) If recovery is attempted, what management strategy should be followed?

Little was known about the fish communities on the NWS at this time and so the CSIRO Division of Fisheries undertook an intensive two year study during 1982 and 1983. As a result of this research a series of alternative hypotheses were developed to explain the observed change in community composition. These were:

i) environmentally-induced changes in the fish community, independent of the fishery;

ii) multiple, independent responses by the fish species to exploitation;

iii) alteration of biological interactions, such as predation and competition, due to fishing;

iv) indirect effects of fishing such as habitat alteration.

While the NWS is influenced by the oceanographic phenomenon known as the El Niño-Southern Oscillation, the precise nature of changes in the community dynamics on the Shelf in response to these events remains unknown and quantitative models based on hypothesis (i) could not be developed. Changes of type (ii) are possible and involve highly density-dependent population parameters (Sainsbury 1987) for which simple models could be developed. Ecological theory offers numerous possible mechanisms which could be used to generate simple models to test hypothesis (iii).

Less was known, however, about the relationship between the fish stocks and their habitat, notably the habitat provided by demersal epibenthic organisms. To examine habitat usage by the major fish types a 35 mm camera system with two illuminating strobes was fitted to the head rope of the trawl gear, and a colour photograph was taken every 24 seconds for the duration of each 30 minute trawl. From these photographs the major habitat types were identified, based on the presence or absence of large epibenthic organisms (> 25 cm along the major axis), and the association between fish and habitat was calculated. The results showed that, in general, *Lutjanus* and *Lethrinus* species occurred predominantly within those habitats which contained large epibenthic organisms, while *Nempiterus* and *Saurida* species favored the open sandy habitats. If these results are indicative of a dependence of certain fish species on certain habitat types, then alteration of the relative amount of each habitat in an area by trawling would alter the fish community composition in that area.

Either separately or together, hypotheses (ii), (iii) or (iv) are able to explain the changes in fish species composition observed on the NWS. However, each of these possible causes of change can have different management implications. If the management objective is to maximise the catch of *Lutjanus* and *Lethrinus* species, and the main cause of the current decline in these species is the loss of large epibenthic habitat (hypothesis iv) and/or trawl-induced changes in competitive/predation interactions (hypothesis iii), then there would be scope for expansion of the trap fishery to replace trawling with an expectation of recovery of the associated habitat and/or a return to the high carrying capacities of these species. On the other hand, if the historical decline in catches of *Lethrinus* and *Lutjanus* species is because these stocks have intrinsically low productivity (hypothesis ii), then the scope for expansion of domestic fisheries of any kind is limited.

3b. Experimental management scheme—1986–1991

To clarify the possible dynamics underlying the multispecies trawl fishery on the NWS, an experimental management scheme was introduced between 1986 and 1991. This scheme consisted of applying three different management regimes on three large areas of the NWS (Figure 1). This management approach is 'actively adaptive' because it includes taking management actions that intentionally try to increase the contrast between control variables. This approach improves identification of the key processes in the dynamics of the managed system. (A more complete description of the application of this approach to the NWS is given in Sainsbury 1991).

CSIRO Division of Fisheries monitored:

1) how the various fish species responded to the changed fishing regimes;

2) the recovery rate of habitats after trawling ceased; and

3) the effect of trawling on habitats.

The monitoring was based on a stratified sampling strategy with 105 randomly located thirty minute trawls undertaken during the same season each year. A photographic transect was conducted for each trawl to monitor the habitat (as discussed earlier). A video camera was used during some trawls to determine the effects of the trawl gear on epibenthic organisms.

Data from the western zone, which was closed to commercial trawling for the duration of the experiment (from Oct. 1985-Figure 1) shows that there was an increase in both the combined populations of *Lutjanus* and *Lethrinus* species (Figure 2a) and the abundance of both large and small epibenthic organisms (Figure 2b). Recovery of the small epibenthic organisms has been fairly rapid, but recovery of larger epibenthic organisms is much slower. This indicates that there will be a considerable time lag after trawling ceases before recovery of large epibenthic organisms is substantial. Within the zone in which commercial trawling continued throughout the experimental period (Figure 1), the opposite trends are observed. Except for the year 1990 (where the catch data are influenced by two very large catches within this zone), there was a steady decrease in the catch rates of *Lutjanus* and *Lethrinus* species (Figure 3a) and an associated decrease in the two sizes of epibenthic organisms (Figure 3b).

These results show a good correlation between the catch rate of the large commercial fish species and the abundance of epibenthic organisms. However, this does not necessarily imply an ecological association, as both the fish and epibenthic populations may be responding separately to the effects of trawling. To test for this possibility, the likelihoods of observing the experimental results described above under alternative models of the resource dynamics were calculated. Three of these models included a response by the fish populations based on changes in interspecific and intraspecific processes alone while a fourth alternative assumes that the carrying capacity of all groups of fish species is determined by the amount of suitable habitat, and habitat abundance is altered by the effects of trawling (for a full description of the models and necessary steps in calculating the likelihoods refer to Sainsbury 1991). Combining the historical catch and effort information with the research data collected on the NWS since 1982, the most likely model of the resource dynamics is that in which the abundance of the major fish groups is limited by the amount of suitable habitat available.

While the effects of trawling on epibenthic habitats may be inferred from the continued decline in such habitats in the areas trawled (and the recovery in the areas closed to trawling), direct video observations were made of the passage of the trawl gear over the sea bed. A video 8 camera was placed in an underwater housing and attached to the headline of the trawl gear. The camera was positioned so that it looked down and back towards the footrope. The field of view encompassed the ground gear and an area of around 3 to 4 metres both in front and behind.

The results of those video observations are given in Table 1 (a complete description and analysis of the observations will be published elsewhere). The fate of epibenthic organisms after impact with the trawl gear remained unknown in over half (52%) of the 393 observations. Where the fate on impact was known (188) epibenthic organisms remained intact in only 10% of the cases, with damage (detachment from the sea-bed) occurring in the other 90% of observations. Bounds on the probability of damage on impact with trawl gear can be calculated by making extreme assumptions about the observations when the fate was unknown. If all observations where the fate is unknown are included in the undamaged category then the probability of damage due to the passage of the trawl is 0.43, whereas if all unknown outcomes are considered to have resulted in damage then the probability of damage is 0.95. Of particular interest in assessing the effect of trawls on the epibenthos is the proportion of damaged items which are caught in the net compared to those that rolled under the net and so would not be caught and seen by ship-board observers. From the video observations only 10% of the damaged organisms were seen to be thrown up and into the net. Hence it would appear that the amount of benthos recovered in a trawl net is only a small component of the benthos that is damaged by the trawl ground gear. Absence of benthos in the recovered net does not necessarily imply that none was damaged

Discussion

The analysis carried out on the North West Shelf has proved very useful in guiding management actions despite the initial levels of uncertainty. Results to date indicate that the composition of the multispecies fish community on the NWS (and possibly other tropical areas of Northern Australia) is at least partially habitat dependent and that historical changes in relative abundance and species composition in this region are at least in part a result of the damage inflicted on the epibenthic habitat by the demersal trawling gear. Furthermore, continued alteration of the demersal habitat due to trawling will probably continue to alter the species composition with increasingly adverse effects on catches of *Lutjanus* and *Lethrinus* species. Recovery of the large epibenthic organisms in the areas which have been closed to trawling appears to be slow.

There appears to be considerable scope for using the 'adaptive' management approach in fisheries research to provide active feed-back between management action and empirical learning. New management measures can and should be introduced in ways that provide information which will guide the choice of long-term management actions. Such approaches can be expected to enhance management for sustainable resource use.

References

Bellamy, Dr. (1842). Guide to the fishmarket. *Quoted in* Street, P. (1961) pp. 34-35 In *Vanishing Animals. Preserving nature's rarities.* Faber and Faber, Great Britian, 232 pp.

Cushing, D.H. (1980). The decline of the herring and the gadoid outburst. *J. Cons. Int. Explor. Mer* **39**, 70–81.

Hempel, G. (1978). North Sea Fisheries and fish stocks - a review of recent changes. *Rapp. Proces-Verb. Reun. Cons. Int. Explor. Mer* **173**, 145–167.

Liu, H. C. and H. L. Lai (1980). Cost-revenue analysis of Taiwanese pair trawlers operating in Australian waters. *Acta Oceanogr. Taiwanica* **8**, 109–140.

Moran, M., J. Jenke, C. Burton and D. Clark (1988). The Western Australian trap and line fishery on the North West Shelf. FIRTA Report 86/28. W.A. Marine Research Laboratories, 79 pp.

Sainsbury, K.J. (1982). The biological management of Australia's multispecies tropical demersal fisheries: a review of problems and some approaches. *CSIRO Marine Laboratory Report* **147**, 16pp.

Sainsbury, K.J. (1987). Assessment and management of the demersal fishery on the continental shelf of northwestern Australia. pp. 465–503 In *Tropical snappers and groupers: biology and fisheries management.* Ed. by J. J. Polovina and S. Ralston. Westview Press, Boulder, CO.

Sainsbury, K.J. (1988). The ecological basis of multispecies fisheries, and management of a demersal fishery in tropical Australia. pp. 349–382 In *Fish population dynamics* (2nd ed.). Ed. by J. A. Gulland. John Wiley and Sons, Chichester and New York.

Sainsbury, K. J. (1991). Application of an experimental approach to management of a tropical multispecies fishery with highly uncertain dynamics. *ICES mar. Sci. Symp.* **193**, 301–320.

Walters, C.J. (1986). *Adaptive management of renewable resources.* Macmillan Press, New York.

Walters, C.J. and R. Hilborn (1976). Adaptive control of fishing systems. *J. Fish. Res. Board Can.* **33**, 145–159.

Walters, C.J. and R. Hilborn (1978). Ecological optimization and adaptive management. *Ann. Rev. Ecol. Syst.* **9**, 157-188.

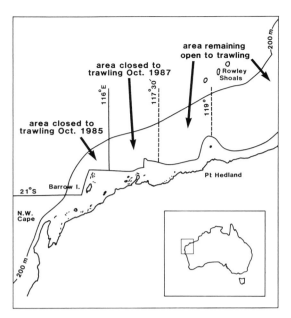

Figure 1. The area on the North West Shelf in which the CSIRO research was conducted. The zoning used during the experimental management between 1986 and 1991 is also shown.

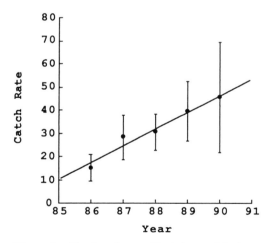

Figure 2a. Total catch rate of *Lethrinus* and *Lutjanus* (kg/30 min) in the zone closed to trawling in October 1985 based on the annual research data. Standard errors and line of best fit are also shown.

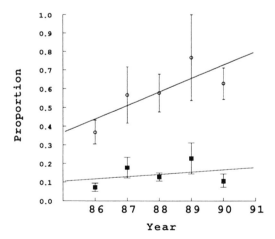

Figure 2b. Proportion of sea-bed with large (closed square) and small (open circle) benthos in the zone closed to trawling in October 1985 based on the annual research data. Standard errors and line of best fit are also shown.

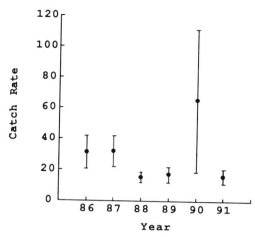

Figure 3a. Total catch rate of *Lethrinus* and *Lutjanus* (kg/30 min) in the zone left open to trawling based on the annual research data. Standard errors and lines of best fit are also shown.

Table 1. Effects of demersal trawling on marine epibenthos

Observed effects due to the passage of the trawl ground gear on epibenthic items larger than the width of the ground gear itself (about 15 cm). These data are based on video observations taken of seven 30 minute demersal trawls.

Type of observation	Number	Percent
No observed effect	19	5
Broken and rolled under net	154	39
Broken and caught in net	15	4
Unknown	205	52
Total observations	393	100

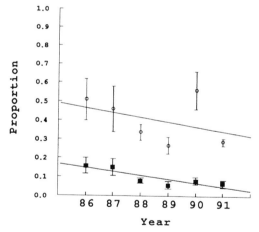

Figure 3b. Proportion of sea-bed with large (closed square) and small (open circle) benthos in the zone left open to trawling based on the annual research data. Standard errors and lines of best fit are also shown.

DISCUSSION OF SESSION 5

Recorded by P.C. Coutin
Marine Science Laboratories
P.O. Box 114
Queenscliff VIC 3225

The presentation of each paper by individual panellists was followed by brief questions and discussion, which are reported here together with the concluding general discussion which was led by the Session Chairperson.

Following *Martin Mallen-Cooper's* panel presentation on changes in the distribution and abundance of freshwater fish due to changes in habitat, Stan Moberly suggested to the Society that a vision of the future of fish habitats is needed. By using the Society's collective wisdom, it should be possible to predict what the human impact on fish habitats would be in the future. Before such an impact occurs, these predictions should be communicated to the decision makers.

Martin Mallen-Cooper responded that the Society could already list the human impacts that have the greatest impact on freshwater fish habitats and proposed that such a list should be compiled during the workshop. It was suggested that the Society should be more proactive in promoting fish habitat issues and that scientists should communicate their existing knowledge.

Referring to water quality and fish movements, Peter Gehrke stated that a great deal was already known about freshwater fish habitats, and significant advances have been made to add to this knowledge. He proposed that scientists should capitalise on this knowledge to prevent further habitat degradation and to actively promote habitat enhancement. For example, fishways should be constructed to allow fish to move upstream above dams and weirs.

Phil Cadwallader cited the Fish Management Plan of the Murray-Darling Commission as a good example of the direction for freshwater fisheries management which involves fish habitat restoration programmes utilising techniques such as fish ladders. However, he pointed out that an essential element was the associated television advertisements produced by the Commission's educational unit which increased public awareness of freshwater habitat issues.

Given sufficient resources, Martin Mallen-Cooper agreed that even a small group of people could change public opinion at grass roots level and suggested that as an outcome of the workshop, the Society should recognise the importance of public education.

John Koehn opened the discussion on *Greg Jenkin's* presentation by commenting that the 70% loss of seagrass and the subsequent decline in commercial fin fish catches—particularly of King George whiting—in Western Port Bay was a classic case of lack of management foresight and lack of publicity about habitat degradation. The loss of seagrass could have been predicted a long time ago if the effects of draining the swampland around Western Port Bay had been thoroughly considered. The subsequent scouring of river beds in the catchments led to massive amounts of sediments entering Western Port Bay which smothered the seagrass beds and elevated the mud banks. Until a few years ago, neither coastal nor catchment managers had recognised the problem. Urgent management action was required to avert the threat of sedimentation from poor land practices in order to maintain the current quality and quantity of freshwater and estuarine fish habitats. Costly habitat restoration programs were required to retrieve aquatic environments from a history of poor land use practices and to allow the quality and quantity of fish habitats to fully recover.

George Paris reiterated the importance of education and publicity about habitat loss. Referring to the proposed re-location of the Coode Island petro-chemical complex to Point Wilson in Port Phillip Bay, Victoria, he suggested that even poor or misleading information on environmental impacts was better than no information in helping to produce the right social and environmental outcome. The current lack of public advice from marine biologists would achieve nothing at all. He proposed that the Society should follow the example of the Royal Australian Ornithological Union by releasing press statements and providing advice on the impacts on the aquatic environment, fish habitats and fisheries, of threatening human activities such as the development of the proposed new site of the petro-chemical complex in Victoria.

Greg Jenkins, however, stressed the need for scientists to state the facts correctly and criticised some environmental lobby groups that made factually incorrect press statements which misled the public. It would be far better for responsible scientific organisations, such as the Society, to release accurate statements at an early stage. Once the wrong facts have been given to the public, it is very difficult to make corrections in the press. Good scientific advice to the media would be really worthwhile.

Julian Pepperell opened discussion on *Robert Campbell's* account of the effects of trawling on marine communities on the Northwest Shelf of Australia by noting that the change in the quality and quantity of habitat on the Northwest shelf has caused changes in fish species composition. He asked the panelist whether the direct fishing mortality caused by Taiwanese pair trawlers, which involved tens of millions of lutjanids and lethrinids, had made a bigger impact on the fish community than the indirect effect of trawl nets through damage to the seabed.

Robert Campbell replied that a number of models and assumptions had been tested by his co-investigator Keith Sainsbury and that the direct impact of fishing mortality on the composition of the fish community was less than the indirect impact of habitat degradation. The models indicated that the habitat-driven response was the most likely factor to have caused the change in the fish populations. Although the recovery of habitat after trawling had ceased appeared to be quite good, it was a slow process and full recovery of the habitat following heavy trawling may take up to twenty years.

Given the effects of trawling on the habitats and fish communities of the Northwest shelf, Russell Reichelt asked whether a similar effect on habitat and fish communities may have occurred in the South East Trawl Fishery. He observed that a map of all the trawl shots made over the last five years showed that there had been a complete coverage over the shelf in south eastern Australia. He suggested that the fish stocks present today represented a 50-100 year shift in population structures that is now being maintained by the effects of trawling on the fish habitat and the fish stocks.

Robert Campbell commented that there had been very heavy trawling on the Northwest shelf and the data indicated a complete coverage by trawlers during the 1970s. While fisheries management is based on a concept of the original fishery there are rarely any supporting data to indicate the composition of the initial habitat and fish communities.

Murray MacDonald opened the *general discussion* by repeating the objectives of Session 5:

- to identify the categories of human impacts on aquatic habitats;

- to assess the scientific evidence for linkages between human activities, habitats and fisheries resources; and
- to identify strategies for demonstrating to the public and decision makers the link between human impacts and the degradation of aquatic habitats.

He suggested that the Society could take a leading role in identifying the full range of human impacts on aquatic habitats and in assessing the quantity and quality of scientific evidence for the linkages between human disturbances of aquatic habitats and the subsequent effect on fish communities. When these tasks have been completed, the Society and individual scientists could then provide guidelines and recommendations on the management of human impacts on fish habitats to decision makers.

The Society could also package the existing habitat impact information in a form suitable for a broader audience of managers, legislators, politicians and the general public. Convincing the public of the need for fish habitat protection and aquatic conservation and providing scientific information in a suitable form would help to create a climate which would allow politicians to make informed decisions about the alternative management objectives and actions that are available to them. He also suggested that the Society could identify the additional information needed by scientists, managers and the public to show more clearly the effects of various human activities on aquatic habitats and therefore fish populations.

A number of categories of human impacts on aquatic habitats were listed :

Freshwater Habitats
- Habitat loss (dams, flow regulation, channelisation)
- Eutrophication (land practices, fertilisers, sewage)
- Sediment deposition (land practices, forestry)
- Salinisation (land practices)
- Contaminants (organic and inorganic)
- Introduction of non-native species (including translocations)

Estuarine Habitats
- Habitat loss (eg. shoreline development, dredging, drainage schemes)
- Changes in salinity (eg. altered river flows, artificial openings to the sea)
- Eutrophication (fertilisers, sewage)
- Sediment deposition (land use practices)
- Impact of fishing methods
- Contaminants (eg. heavy metals and hydrocarbons)
- Introduction/translocation of exotic species (eg. oriental goby, toxic algae)

Marine Habitats
- Impact of fishing methods on the seabed and benthic community (trawling/dredging)
- Hydrocarbons (oil contamination)
- Waste and debris (eg. ocean dumping)
- Effluent disposal

Global Effects on Fish Habitats
- Climate change / the Greenhouse effect
- Changes in sea level / rainfall
- Ozone depletion

Referring back to the Habitat Conservation Program of the US National Marine Fisheries Service, Murray MacDonald outlined the threats to living marine resources and habitats that had been identified and ranked on a regional and national basis in the USA. Both managers and scientists had developed strategies to address these major habitat threats, had prioritised the implementation of those strategies within the constraints of existing resources, and had also sought additional resources. Key information required to determine each human impact on fish habitats has been identified and research tasks to address specific information requirements are undertaken in priority order. Information so obtained is used by government agencies to guide and control the human activities which have a major impact on fish habitats. Murray MacDonald suggested that perhaps Government agencies in Australia could adopt a similar approach.

Stan Moberly cautioned that in spite of the progress that had been made there was still a major requirement in the USA for more money to address fish habitat issues. The National Fish and Wildlife Foundation had recently recommended to the US government that in addition to the $8 million in the NMFS's habitat budget, an additional $12 million dollars should be provided to make a total of $20 million. However, the House of Representatives voted not to increase the level of funding to address fish habitat issues, and the Senate voted to increase funding by only $2 million. When the House and the Senate conferred on this matter they finally voted to allocate no extra funds at all. So no matter what strategy is developed, and no matter how much research on the impacts of human activities on fish habitats is done, it achieves nothing until the Government gives the issue a high priority and allocates sufficient funds to tackle the problems of habitat degradation and loss. Increasing numbers of people in the USA are becoming aware of habitat issues, but funds are not being made available to address these issues.

Murray MacDonald asked how the Society could convey the habitat message in Australia in a manner that would convince politicians to give habitat issues a high priority. Mick Olsen wondered whether the Society could support already existing public education programmes like that of the Murray Darling Basin Commission.

Phil Cadwallader agreed that the Murray-Darling Basin Commission had recognised some of the key factors that caused habitat degradation and had developed a management plan to tackle habitat issues. More importantly, the management plan had been "sold" to the public using television advertisements to ensure that there was a high level of understanding and acceptance. Land care programmes such as re-afforestation and re-vegetation have also been well supported due to effective publicity campaigns which have been developed as part of a national resource management strategy. However, he agreed with Stan Moberly that the most important requirement in addressing habitat degradation and loss is money. There is a need to convince the public of the seriousness of these issues, and scientists should adopt more of an educational or extension role, like writing popular articles and using television to get the message across. It is necessary to get away from the ivory tower mentality and from closed academic circles. If scientists leave the job of communicating habitat issues to media journalists or to public relations officers in Government departments, the message will invariably be wrong. The facts really need to be told by scientists. Martin Mallen-Cooper agreed that communication and community education need to be given priority by the Society and that the necessary funds need to be obtained.

Rob Lewis identified the need to communicate clear and simple messages to the public, and cited as an example the successful national publicity campaign to reduce littering. He claimed that people litter far less today because of the successful poster campaign and a short, simple television campaign—not because scientists put forward piles of detailed environmental impact assessments into the public domain. With simple messages, children and adults could be convinced not to cause a littering problem. He suggested that the Society should consider a poster campaign on the fish habitat issue, with short snappy phrases that will influence people over a period of time, rather than offering detailed scientific advice which most people are not going to accept or even understand.

Murray MacDonald agreed that it was vital to pitch the message at the right level depending on the audience. He asked for comments on the role of scientists in providing simple environmental messages for public awareness campaigns, and on whether scientists actually need to collect detailed scientific information for such purposes. Rob Lewis reiterated that publicity campaigns needed to be consistent, simple and widespread, but stressed that it was vital that scientific research should continue to be conducted to provide an authoritative factual basis for the message to the public.

Margaret Shepherd added that a financial incentive, such as fines or an environmental levy, was a very effective way of rapidly bringing environmental issues to public attention. Using the example of the environmental levy imposed on water rates in NSW, she pointed out that the public was very quick to demand action when rubbish from the deepwater sewage outfalls appeared on Cronulla beach. The public had refused to pay the rates until the beaches were cleaned. She suggested that an environmental levy was a good way to raise public awareness and to obtain funds for research. Rob Lewis agreed that the public will want value for money if Governments decide to apply an environmental levy. However, the community will not demand an environmental levy without a proper marketing program which promotes habitat issues and convinces them of the need for a levy.

Bob O'Boyle returned to the problem of funding habitat research and gave a Canadian example of raising research funds from offshore oil exploration that developed in the 1970s. Initially, the Government of Canada established very strict guidelines in relation to the Environmental Impact Statements (EIS) for offshore oil exploration which involved individual companies funding large scale ichthyoplankton surveys, full scale demersal fish surveys, and oceanography. However all the data that were generated by those surveys were confidential, and all the other oil companies had to repeat the surveys if they wanted to obtain their own information. Consequently, the oil industry established a national fund, known as "The environmental studies revolving fund" (ESRF). All the oil companies put in a small percentage of their profits, thus generating millions of research dollars. A committee consisting of Government and oil industry representatives was set up to administer the tendering for all oil exploration projects requiring Environmental Impact Statements. The ESRF funds were thus used to collect scientific information which could be accessed by the entire oil industry. This approach solved the funding problem as it basically made use of the 'user pays' system - a concept which may be applied to aquaculture and other natural resource-based industries.

Peter Jackson agreed that the Society had to act now and that it could not wait for more research results. The Society could already list the threats to Australian freshwater habitats at least and, more importantly, could identify some of the broad strategies required to deal with those threats. These strategies should be communicated to politicians and managers as quickly as possible. In Queensland the state Government is frequently being asked how to protect freshwater habitat, and scientists should provide up-to-date information in response to such requests. However, it is important to recognise that gaps in our knowledge of human impacts on freshwater fish habitats still exist. Peter Jackson proposed that the Society should identify those gaps and put forward a research strategy to collect the missing information.

Jim Puckridge believed that it is the Society's responsibility to communicate available information on the impacts of human activities on fish habitats. He suggested that at a future Society workshop a session should be devoted to explaining how scientists can more effectively communicate with the media, attract media attention, and provide a coherent and effective message. David Smith identified two types of information that scientists should provide. The first type was detailed scientific advice which identified key issues and options for policy makers and set priorities for research.

The second type of information was the short, simple message to the public through the media or a poster campaign such as "Fishers Involved in Saving Habitat".

Referring to the demand for a ban on commercial scallop dredging in Port Phillip Bay, Peter Young cautioned against providing advice in the absence of supporting scientific information. He stated that, unlike the apparent situation for freshwater habitats, relatively little was known about the impacts of human activities on fish habitats in marine and estuarine environments. Russell Reichelt re-emphasised the need for solid scientific information to back up the simple messages of a publicity campaign. He expressed concerned at the apparent lack of compiled information that showed the total loss or degradation of aquatic habitats on a national or statewide basis due to human activities such as canal estate development, swamp drainage schemes, and pollution of estuaries, mangroves and seagrass beds. This broad scale information was needed to convey the seriousness of the habitat issue to the media and the public.

Julian Pepperell, as President of the Society, considered that it was well within the scope of the Society's charter to put together some good graphic material for the media and general public that showed the loss and degradation of aquatic habitat in Australia. He proposed that scientists in the Society with good computer graphic skills could produce, for example, maps showing the decline in seagrass beds and mangrove stands around the entire Australian coastline. Once the material is compiled, the next problem is to broadcast the message to the general public. However, attracting the media's attention on such issues is difficult. Television is the best way of conveying the message to a wider audience, but this would require Government funds. Another problem is that many individuals do not perceive any obvious personal responsibility for environmental impacts on fish habitats. Projects such as construction of dams, de-forestation, sewage disposal, and land use practices which cause the gradual siltation of seagrass beds, usually represent the combined activities and/or requirements of many people in the private and public sectors of the community, and their environmental impacts cannot be attributed directly to specific individuals.

John Glaister noted that there are quite a few State and Federal government agencies and non-government organisations that are already providing considerable resources for public education on fish habitats. For example, Ocean Watch, the Queensland Commercial Fishermen's Organisation and Queensland Fisheries have all conducted public awareness programs on fish habitat issues. He suggested that the Society should coordinate a national campaign and capitalise on these efforts. He proposed that the Society should establish a Threatened Habitat Group, consisting of key people from State and Federal organisations, to collate information on the status of aquatic habitats. Murray MacDonald added that to carry out these tasks more quickly and efficiently it could be appropriate to combine the resources of the Society with other professional societies like the Australian Marine Sciences Association and the Australian Society for Limnology.

Stan Moberly stressed that the Society should focus on transferring the knowledge that we have already in simple and straightfoward terms. If there are unanswered questions, these should be used to identify and focus research needed. However, this need for further research should not be the main message scientists send to the public. We need to convey to the public what we already know so that this information can be used to help make decisions the community faces now. For example, it is not worth arguing to the public exactly how bad sewage is for fish habitats; the public just needs to know that it *is* bad and does not improve fish habitats. When someone asks "Just how bad is the sewage?", we should convey the message that "it doesn't taste very good and its not good for fish!". Another example is the reduction of river flows through water abstraction or storage in dams. The public doesn't need to hear our scientific debates over *how much* water flow needs to be maintained, but rather they need to know that if we expect fish to migrate upstream and pass through fishways around dam walls, there has to be *sufficient* water flow to accommodate that movement. Simple messages are required for the public, such as :

"Dumping dredge spoil over seagrass beds does not improve fish habitats".

"Fish need water to migrate upstream".

"If you want seafood to taste good, don't dump sewage in the water".

"If you want to swim on nice beaches don't dump sewage in the water".

These are the kind of messages that raise public awareness and consequently raise the profile of fish habitat, water quality and water quantity issues. It should be a simple process to apply the successful media campaign on littering to a simple message on fish habitat conservation. In the USA the campaign on marine debris swept the nation as a result of an aquatic education programme which pointed out the problem and suggested solutions, but didn't get into a debate about how much marine debris is bad. The same approach could be taken in Australia.

Murray MacDonald concluded the discussion and urged the membership and the executive of the Society to implement the ideas that had been put forward.

SESSION 6

ALTERNATIVE USES (ECONOMIC, SOCIAL AND POLITICAL)— KEY VALUES

Session Chairperson: R. Reichelt
Session Panellists: D.A. Pollard
A.J. Staniford
Rapporteur: J. Robertson

CHAIRPERSON'S INTRODUCTION

R.E. Reichelt
Bureau of Rural Resources
GPO Box 858
Canberra ACT 2601

Use of a habitat means different things to different "users". In some countries there are those for whom a shallow coastal coral reef is of more use as a landing site for a heavy military vehicle. In Australia the same reef might be regarded as an ideal site to dredge a stable speedboat harbour or build a canal estate. There would be others who would consider the best use of the reef to leave it as it is.

The Australian public (and media) are becoming increasingly aware of the importance of our inland waterways and "Injured Coastline". However, both fisheries scientists and the public are still desperately short of hard information about the role of particular habitats in sustaining both fish populations and other living resources. For example, an interviewer recently asked me whether the destruction of any particular mangrove stand near Coffs Harbour could explain the disappearance of younger age classes of gemfish from the winter spawning run.

When conflicts between users of a common resource occur in Australia they are handled in various ways, including publicity campaigns, Resource Assessment Commission Inquiries, legal challenges and civil disobedience. All too often the conflicts are treated with confusion, apathy and political procrastination. Although the complex jurisdictional arrangements can exacerbate the situation, biologists are frequently unable to provide the information called for by resource economists in evaluating particular conflicts.

Analysts often maintain different world views. Economists sometimes regard the universe as an economic system within which ecology is one component, many others regard economics as one component of our ecosystem, and a third group regard economics and ecology as separate dimensions within our human niche.

From a fish biologist's standpoint the large problems are:

- defining the environmental impact caused by, or likely to be caused by, an alternative use;

- recommending options for controlling the degree of the impact (or effect);

- developing precise and comprehensible (not necessarily quantitative) assessments of the risks associated with alternative uses.

Outside the usual brief for fish biologists lie the realms of bioeconomic risk assessment, resource pricing and establishing the full range of alternative reasons that people may want to use fish habitats. Yet the community expects fish biologists to provide a sound input to analyses in these other areas.

Fish biologists should aim to have an effective input to both the policy making process and the use of scientific results in decision making concerned with regulating the use of fish habitats by a broad spectrum of users. Our panel speakers will spark debate on these broad issues and the approaches that fish biologists could take to achieve this aim.

MAXIMISING THE POTENTIAL FOR BOTH SUSTAINABLE FISHERIES AND ALTERNATIVE USES OF FISH HABITAT THROUGH MARINE HARVEST REFUGIA

D.A. Pollard
Fisheries Research Institute
P.O. Box 21
Cronulla NSW 2230

The main aim of this session of the Workshop, as I interpret it, is to stimulate and provoke discussion on the management of "fish habitat" in relation to maximising the compatibility of alternative non-consumptive uses with the consumptive uses of fisheries resource harvest.

As the term "fish habitat" is a very broad one, and could include most aquatic ecosystems, in introducing this discussion I will concentrate mainly on the marine/estuarine system. Although the problems of management of freshwater/inland aquatic systems are in many ways both qualitatively and quantitatively different, any discussion from the floor of the applicability of what I have to say in relation to freshwater ecosystems would also be most welcome.

I don't intend to go into the details of the relative importances of different types of fish habitats, which are discussed in other sessions of the workshop, though such shallow water inshore marine and estuarine nursery habitats as seagrass meadows, mangrove swamps, and rocky and coral reefs could be kept in mind because of their ecological importance and vulnerability. Neither do I intend to go into the details of the many and varied alternative human uses of aquatic ecosystems, though I would like to emphasise that we should not regard the destruction of fish habitat through such activities as dredging and reclamation for foreshore development as being legitimate "uses" in this context. The sort of acceptable alternative non-consumptive uses to which I will be referring could be broadly grouped as recreational, educational, scientific and aesthetic—the demands for all of which might be expected to increase with increasing affluence, leisure time and tourism in the not too distant future.

What I would like to do to stimulate discussion is to consider a future *scenario* in which demands for these alternative uses, and also the demand for fisheries products, have greatly increased, and to consider an appropriate management regime to address the inevitable ecological pressures and use-conflicts which will arise.

The marine/estuarine system has, of course, traditionally been regarded as a "commons" (but see footnote below*), and until relatively recently a generally open slather approach to its usage has been dominant and almost universal. Needless to say, this "let her rip" *laissez faire* approach cannot be allowed to continue, and

---

* "Despite widespread usage of the term 'commons', our current institutional structure for ocean management does not meet one critical criterion for common property: community control. The tragedy of the ocean is not the tragedy of the commons, but the tragedy of overuse. Overuse may result from fragmented and ineffective ownership. Overuse may also result from short-term profit taking by private owners. It is a red herring to link overuse to common ownership" (Hanna 1990).

viewing the system from a multiple-use perspective, the compatibility and conflicts of the many different potential uses urgently need to be considered in terms of both ecological sustainability and user equity.

In this regard we need to consider the most appropriate management regime to achieve these basic aims of sustainability and equity.

Overall, I would consider that the maintenance of ecosystem biodiversity, in terms of both the sustainability of the fish (i.e. the stocks of renewable, and thus potentially harvestable, resources) and the preservation of their habitats (i.e. both the biotic and non-biotic components of their environments), should be the most critical and primary management objective. While it obviously needs to be considered concurrently with the above, effective management for equity of use allocation amongst the many different competing user groups can only really be achieved once the maintenance of biodiversity has been seriously addressed.

At present, the management of "fish" and their habitats in most areas of the marine/estuarine system involves the piecemeal management (sometimes successful, but often not) of individual species, stocks, habitats and uses, carried out by a plethora of different and often competing management authorities, each often working within its own narrow and conflicting legislative and jurisdictional framework. While this is obviously an improvement over the previous practically unmanaged state, in my view this approach cannot hope to assure the maintenance of biodiversity, and thus equity of use, in the longer term. What I therefore suggest is needed is a much more holistic approach to aquatic ecosystem management.

In listening to the papers presented and the subsequent discussion over the past day and a half of this Workshop, I have started to get the impression that, on this habitat issue in general, maybe most of us have been, like Nero, fiddling (in our case with our nets and our computers) while Rome burns.

Over the many years we have all spent seeking the details of whether and why juvenile banana or tiger prawns preferred seagrasses or mangroves, or whether and why juvenile blackfish or leatherjackets preferred *Zostera* or *Posidonia*, vast areas of these valuable fish habitats have been degraded and destroyed. Now, I'm not suggesting that we stop carrying out this type of detailed ecological research, but we certainly shouldn't wait to do something practical until we've found out all the answers.

After nearly 30 years in this field, to me, two things have become increasingly obvious: the first is that we are never going to understand very much about how aquatic ecosystems really work (e.g. exactly why do those baby banana or tiger prawns prefer seagrasses or mangroves?); and the second is that, in the longer term, we are not going to improve much on, or be able to replace, Nature's fine work in creating the fish habitats that we already have.

The natural habitat we see around us—for fish and everything else—is about as good as it's ever going to get. For the continued maintenance of wild stocks, there is really only one obvious and immediate thing to do, and that is to preserve as much as possible of that which remains in its natural state, while we still have the chance. If we do so, we can continue harvesting part of its productivity for our own purposes in perpetuity.

There is thus not such an urgent need to understand the details of exactly how "fish habitat" works—that it does work has already been taken care of for us. To reiterate, what *is* most urgent is to preserve as much of it as possible in its natural intact state. Thus, maybe one of our most vital tasks as an informed scientific Society should be to preach this message much more strongly out there in the community in general—not only to the fishermen, engineers, planners, policy makers, etc., but also to the committed (though sometimes uninformed) 'greenies', the school kids, and all the rest.

In providing some food for serious thought then, I'll be provocative and suggest that the *only long term solution* may lie in the total protection from consumptive uses of very large areas of the marine and estuarine environment. The establishment of such extensive "marine habitat refugia", coupled with the use where necessary of more conventional but less holistic management measures in the surrounding fished areas, would help to replenish the fishable stocks in these areas through protection of both adult spawning stocks and juveniles in their nursery habitats. Apart from providing continuous recruitment of harvestable resources to fished areas, such marine habitat refugia would also thus be available for those alternative and relatively non-consumptive recreational, educational, scientific and aesthetic uses mentioned earlier, with existing conflicts between the two being greatly reduced.

Although it only affects the waters adjacent to one State, the Great Barrier Reef Marine Park provides a model which demonstrates that a multiple-use management approach based on the general principles of ecological sustainability and user equity can successfully operate within a suitable framework of Commonwealth/State co-operation. I suggest, however, that with increased usage the declaration of much larger areas of totally protected habitat as marine harvest refugia will be necessary in order to maintain biodiversity in the longer term, in not only the waters of the Great Barrier Reef but of Australia in general.

Details of how such refugia may work, and assessments of their effectiveness, may be found in the references which follow.

References (not cited in text)

Bohnsack, J.A. (1990). The Potential of Marine Fishery Reserves for Reef Fish Management in the U.S. Southern Atlantic. *NOAA Technical Memorandum NMFS-SEFC-261*, Miami, 40 pp.

Davis, G.E. (1981). On the role of underwater parks and sanctuaries in the management of coastal resources in the south-eastern United States. *Environmental Conservation.* **8**, 67–70.

Davis, G.E. (1989). Designated Harvest Refugia: the next stage of marine fishery management in California. *CalCOFI Rep. No.* **30**, 53–58.

Hanna, S.S. (1990). The Eighteenth Century English Commons: a Model for Ocean Management. *Ocean and Shoreline Management* **14**, 155–172.

Roberts, C.M. and V.C. Polunin (1991). Are marine reserves effective in management of reef fisheries? *Reviews in Fish Biology and Fisheries* **1** (1), 65–91.

ALTERNATIVE USES OF AQUATIC HABITATS: AN ECONOMIC PERSPECTIVE

A.J. Staniford

Office of Energy Planning
30 Wakefield St
Adelaide SA 5000

Abstract

Alternative uses of aquatic habitats are classified according to human participation and resource depletion. The classification assists to clarify the impact of human activity on the resource. Each use places a demand on the habitat that may be compatible with other uses or mutually exclusive. The key question from a management viewpoint is how to meet these alternative demands.

Three options are discussed; the market approach, the regulatory approach and the value-added approach. It is demonstrated that the value-added approach is the preferred option. The capital value of the habitat is explicitly recognised with this approach. It provides a strong basis for developing strategic, forward-looking, consultative management programs that will effectively meet the alternative demands placed on the habitat.

Aquatic habitat management in Australia is moving towards a value-added approach. The rate of progress, however, needs to be accelerated. Actions required to progress this approach are proposed.

Introduction

Aquatic habitats provide a range of alternative uses to society with the potential to improve welfare. However, there are many examples of aquatic habitats where the demands placed on the resource have led to overexploition and permanent resource depletion of one or more of the organisms living in the habitat. In order to prevent overexploitation, the various uses of the habitat must be managed to ensure that the associated impacts are sustainable.

In this paper, I provide an economic perspective to the alternative uses of aquatic habitats. The term habitat is interpreted in its broadest sense to be a region of space where living organisms interact. Alternative uses of habitats are categorised into three groups and alternative management approaches are contrasted.

Alternative uses of aquatic habitats

Aquatic habitats are used for a variety of purposes, ranging from providing leisure activities for society, to being used as a means to generate economic wealth (e.g. fishing). In this paper, I categorise alternative use of habitats into three groups (Table 1).

Category 1 is defined as actions that are both participative and depletive. Participation refers to activities in which there is direct interaction of human activity with the habitat. The level of depletion is defined as the extent to which human activity reduces the quality of the asset. Depletion may be temporary or permanent. Category 1 includes commercial and recreational fishing. It also includes activities such as waste disposal and the use of aquatic habitats for commercial and industrial processes.

Category 2 is defined as participative non-depletive uses. In these, humans interact directly with the habitat but do not deplete the asset. Activities include commercial shipping, recreational boating, surfing, swimming, snorkelling, diving etc. and the use of habitat for aesthetic purposes.

Category 3 is defined as non-participative non-depletive uses. Even if individuals never visit or deplete a habitat, the fact that the resource exists may in itself be beneficial (Kratilla and Fisher 1975). Mitchell and Carson (1989) classify existence (or non-use) benefits into two types - vicarious consumption and stewardship. Vicarious consumption refers to benefits gained from knowing that others are using the resource. In the case of an aquatic habitat, an individual may derive benefit from knowing that members of his family, friends and the general public are able to use the habitat to pursue activities, such as those listed in Categories 1 and 2, that will improve their utility. Stewardship values involve the desire to see a public resource used in a responsible manner and conserved for future generations. For example, individuals may derive benefit from knowing they can pass on a specific aquatic habitat to the next generation even if they do not personally visit the site (bequest value). Also individuals may consider that some habitats are intrinsically valuable to society and should be preserved regardless of whether they will ever be used by anyone (inherent value).

Each alternative use identified in Table 1 places a demand on the habitat. The nature of the demands arising from each use is also indicated in the Table. Category 1 produces a demand for depletion (a reduction in the quality of the asset) *and* space (to enable human interaction with the habitat). Because Category 2 is non-depletive, the demand created is only for space. Category 3 creates a demand for sustainability; provided the resource is sustained, the demand for Category 3 uses can be met.

Some of these demands are compatible with other uses, while others are incompatible. The key objective of management is to ensure that the aquatic habitat is used efficiently, to meet where possible society's demands for the use of the habitat, taking into account the finite nature of the resource. If alternative demands cannot be met on a sustained basis, management should ensure that the alternative uses are rationed to maximise the contribution of the resource to the welfare of society.

Management options

Alternative management options available to allocate the services derived from an aquatic habitat to competing demands can be classified into three categories.

1. The unregulated market approach

Economic theory demonstrates that the benefits (social welfare) from using a resource will be maximised when potential alternative uses are allocated according to the equilibrium of a competitive market system (e.g. Gravelle and Rees 1981). In recent years, governments throughout the world have been increasingly using market mechanisms to achieve efficient use of natural resources in the economy.

Markets exist for some of the products that originate from aquatic habitats (e.g. fish, tourism) and for some of the inputs required to enable people to derive benefit from the resource (e.g. recreational fishing and boat supplies). One management option is to allow alternative uses of the resource to be determined according to the market outcome.

However, there is a large body of evidence available demonstrating that the operation of an unregulated market in this context will not produce a socially efficient outcome. Aquatic habitats have common property (Gordon 1954) and public good characteristics (Cornes and Sandler 1986) that ensure that the market outcome will not be optimal. The failure of the market to produce an optimal outcome is referred to as market failure (Tietenberg 1988).

2. The regulatory approach

Intervention by the government is often advocated to correct situations of market failure (provided the cost of intervention does not exceed the benefit). Intervention is usually achieved through implementation of a regulatory approach. With a regulatory approach, governments recognise that the demands placed on habitats from alternative uses sometimes exceed the capacity at which they can be supplied on a sustainable basis. They therefore intervene to introduce controls on either participation (e.g. limited entry) or depletion (e.g. bag limits or quotas) to prevent overexploitation.

Using the terminology of this session, the key value implicit in application of the regulatory approach is preservation of the habitat and interacting organisms. Regulations are developed and implemented and the effectiveness of the regulations is assessed according to the degree of success in maintaining the habitat.

The regulatory management approach can be extremely effective at preventing overexploitation of individual species by restricting the level of exploitation of the habitat. However, it is characterised by two major deficiencies: it is inflexible and inefficient. Once regulations are enacted, they are often difficult to alter quickly should circumstances change. If a regulatory approach is to be used, effort should be directed to ensuring that the regulations enacted are as flexible as possible to enable managers to respond to unforeseen perturbations of the aquatic habitat.

Even if regulations are flexible, a regulatory approach will encourage users of the habitat to pursue strategies that minimise the adverse impact of the regulations on their activities. This effect is clearly demonstrated through analyses of the impact of regulations on fishing effort in commercial fisheries e.g. Anderson (1985).

Fishers respond to the regulations by substituting non-regulated inputs for those that are regulated. This action reduces the effectiveness with which the regulations protect the stock, implying that the severity of the regulations will eventually have to be increased to maintain the resource at the desired level. More importantly, input substitution by fishers increases the cost of harvesting the fish. Potential profits that could be earned from the fishery are thus reduced. The reduction in profits reduces the economic contribution of the fishery to society.

A further characteristic of the regulatory approach is that regulations are typically introduced after a problem becomes apparent. Thus, even if industry is consulted before implementation of the regulations, the urgency of the problem to be solved often produces political actions and conflict amongst participants. The regulations implemented are often a compromise and oriented towards solving the current problem.

3. The value-added approach

A value-added approach recognises that aquatic habitat is a natural asset that has a capital value. The value of the asset is the sum of the benefits derived by all current and future users of the resource. A value-added approach to management seeks to maximise the total value of the asset.

In contrast to the regulatory approach, the key value is the asset value, not the need to preserve the resource. Due to the market failures associated with aquatic habitats, a value-added approach will include regulations. This may involve input controls and/or output controls. However, the focus of the regulations will be fundamentally different from the regulatory approach. With the value-added approach, emphasis is redirected from preventing overexploitation or protecting the resource, to looking for ways to improve value.

Asset value could be improved by:

- varying the level of exploitation of the habitat (e.g. reducing fishing effort);

- changing access rules to alter the the way in which the resource is allocated between existing participants; and
- encouraging new uses that develop the habitat.

The value-added approach will by definition guarantee protection of the resource. Actions that reduce the quality of the asset will reduce the value. Thus actions that lead to resource depletion that is not sustainable are inconsistent with a value-added approach.

The value-added approach is preferred to the regulatory approach because, not only does it protect the resource, it also facilitates development within biological and economic constraints. Key characteristics of the value-added approach are contrasted with the regulatory approach in Table 2. A key distinction is that the value-added approach is forward looking and anticipative. As a consequence, regulations are flexible and strategic.

Where does Australia fit into this classification?

Aquatic habitat management in Australia is moving towards a value-added approach. For example introduction of individual transferable quotes (ITQ's) in the tuna fishery has assisted to increase the value of the fishery (Geen and Naylor 1989). However, most managers would also agree that there are many deficiencies with the present system. The challenge for the 1990s is to build on the progress to date and to strive for a full value-added approach.

What is required to implement the value-added approach?

1. Habitat managers need to develop a value-added mission. They need to think in terms of values and recognise that in most habitats, the present use pattern will not be maximising the value of the resource. Managers need to search for innovative methods to increase value through changing current use patterns.

2. More research effort needs to be directed towards determining current and potential values of habitats. From an economic perspective this will involve gaining competence in measuring the values derived from alternative uses of habitats e.g. recreational and commercial uses. Biological research needs to provide a sound understanding of the dynamics of the resource and also the impact of human activities on the quality of the asset (which is related to value).

3. Use available technology to improve habitat management efficiency. There are unlikely to be new instruments discovered to control human activity in aquatic habitats. However, technological advances may improve the effectiveness of existing methods whether they be input or output controls

4. There is large scope for institutional reform to improve the effectiveness of aquatic management. Institutions need to be flexible, adaptable, depoliticised, strategic, consultative, anticipative and conflict reducing. The Commonwealth has considered this issue and chosen a statutory authority. What should the States do? Options include:

    - a statutory authority like the Commonwealth's;
    - a Fisheries Board consisting of government and industry representatives; and
    - an independent Fisheries Commission conducting periodic audits of fish habitats and reporting to the Government (analogous to the role performed by the Industry Commission in economic development).

Buchanan (1987) argues that it may be more effective to improve economic welfare by changing the institutions and rules governing the use of resources to obtain a result that more accurately reflects the preferences of stakeholders than to develop strategies (techniques) to improve efficiency within the current rules. In some aquatic habitats, there may be scope to develop common property management regimes rather than continue with management programs that seek to strengthen individual participants' property rights.

5. Information and education of industry and the public are essential elements of a value-added approach. Effective debate on alternative options must be based on accurate information. This must be disseminated to all stakeholders.

6. Increased emphasis needs to be given to the process of fisheries management i.e. consulting with industry, developing, planning and recommending policy. A characteristic of the value-added approach is shared ownership of the problem and the solution. The fisheries management development process will to a large extent determine the extent to which problems and solutions are owned by stakeholders.

Summary

Alternative uses of aquatic habitats place demands on the resource that sometimes exceed the capacity of the habitat to meet the demands. This leads to overexploitation, permanent resource depletion and reduced economic welfare.

Management authorities have typically responded to the potential problem of overexploitation by introducing a regulatory management approach that protects the resource. This approach can be extremely effective in maintaining the resource. However, a regulatory approach is often inflexible and inefficient. Moreover, it may restrict the extent to which the resource contributes to economic welfare. Reduced welfare is expressed through a reduced value of the habitat relative to what it could be under an alternative management arrangement.

A preferred option is to introduce a value-added approach. Rather than just protecting the resource, a value-added approach seeks to develop the resource and maximise the value of the asset through varying exploitation levels, and utilisation of the resource by competing participants. Development opportunities are exploited so that management becomes forward looking and strategic. Implementation of a value-added approach will increase the contribution of the habitat to the welfare of society.

References

Anderson, L.G. (1985). Potential economic benefits from gear restrictions and licence limitation in fisheries regulation. *Land Economics* **61(4)**, 409–418.

Buchanan, J.M. (1987). The constitution of economic policy. *The American Economic Review* **77(3)**, 243–250.

Cornes, R. and T. Sandler (1984). Easy riders, joint production and public goods. *Economic Journal* **94**, 580–598.

Geen, G. and M. Naylor (1989). Individual Transferable Quotas and the Southern Bluefin Tuna Fishery: Economic Impact. Occasional Paper 105, Australian Bureau of Agricultural and Resource Economics, AGPS, Canberra.

Gordon, H.S. (1954). The economic theory of a common property resource: the fishery. *Journal of Political Economy* **62**, 124–142.

Gravelle, H. and R. Rees (1981). *Microeconomics*. Longman, New York.

Kratilla, J.V. and A.C. Fisher (1975). *The Economics of Natural Environments: Studies in the valuation of Commodity and Amenity Resources*. James Hopkins University Press for Resources In the Future, Baltimore.

Mitchell, R.C. and R.T. Carson (1989). *Using Surveys to Value Public Goods: The Contingent Valuation Method*. Resources In the Future, Washington, D.C.

Tietenberg, T. (1988). *Environmental and Natural Resource Economics*, Scott, Foreman and Company, Illinois.

Table 1. Classification of uses

Category	Description	Nature of demand
1	Participtive and depletive	Depletion Space
2	Participative and non-depletive	Space
3	Non-participative and non-depletive	Sustainability

Table 2. Characteristics of the regulatory and value-added approaches

Regulatory approach	Value-added approach
Key value is the stock	Key value is the value of the asset
Protects the stock	Develops the resource
Current problem oriented	Forward looking and anticipative
Inflexible	Flexible and strategic
Consultative	Highly consultative
Political	Depoliticised
Conflict	Reduced conflict

DISCUSSION OF SESSION 6

Recorded by J. Robertson
GBRMPA
P.O. Box 1379
Townsville QLD 4810

Each panel presentation was followed by a time for questions, after which the session was opened for more general discussion.

Following *David Pollard's* panel presentation, John Koehn expressed some doubt as to whether conservation areas were practical in freshwater systems as it would require placing a national park around the whole catchment area. He suggested that, apart from small areas such as small catchments or headwaters, a more integrated approach is required in which the key factors causing detrimental effects are identified. In addition, limits or tolerances should be established in order for amelioration to occur. David Pollard agreed that freshwater systems require a different management approach but nevertheless the reservation approach was still worth considering in some cases.

Bryan Pierce supported the comments of John Koehn in that protected areas may be applicable where a fish species may have a limited distribution or range but not in open river systems such as in South Australia. He suggested that we should be considering, as another alternative, multiple uses, which are really multiple *abuses* of the system, and developing ways in which these uses are not so destructive. David Pollard pointed out that the fish stocks in the area surrounding the refugia need to be managed on the traditional fisheries basis. The insurance policy is the large refuge that allows fisheries managers a 'second go' if initial management plans in the surrounding areas are ineffective.

Peter Young asked how the concept of multiple use systems would fit into the Western Australian Water Board's proposed sewage outfall and the accompanying research programme being developed to evaluate the assimilative capacity of that ecosystem. David Pollard replied that sewage input should be considered as one component of multiple use. Sewage outfalls should be located where they have minimal impact on the refugia.

Jim Puckridge pointed out that multiple use is not a panacea. It can be very expensive and require more substantial expertise than single use refugia. He agreed with John Koehn that an integrated management approach is required in freshwater systems but there is still a case for reserving important or sensitive areas such as highly developed flood plain areas or internal deltas within a multiple use catchment system. David Pollard emphasised the need to be aware of the increasing economic value of non-consumptive uses eg. tourism on the Great Barrier Reef, and start catering for them now.

Following *Andrew Staniford's* panel presentation, Peter Young enquired whether a situation may arise under a value-added approach, where no fishing at all would occur but rather areas being solely reserved for tourist usage. Andrew Staniford agreed that was a possible outcome but suspected it was improbable until there is an alternative source of food fish eg. from aquaculture.

Murray MacDonald endorsed the value-added approach and quoted figures placed on coastal wetlands by the National Marine Fisheries Service in the USA. He queried, however, to what extent trying to put some kind of economic value on these resources is going to compromise the ability to equitably allocate the uses. Andrew Staniford replied that allocation is difficult but requires an understanding of the values of a particular asset in its various uses; and how that resource is shared between the alternative options that are available for that resource.

John Glaister commented that basic values are part of a framework upon which people react to different management alternatives, and values should be expressed in more than just dollars and cents. It was recognised at the Athens conference that for better fisheries management we need institutions that are more diverse, creative and responsive to ecological and social complexity and uncertainty. This requires co-management institutions that incorporate interest, knowledge and wisdom of all resource users not just fishermen. Andrew Staniford responded that at present, as fisheries economists, we haven't enough expertise to place a dollar value to resources with any accuracy but it certainly helps to think in these terms or to identify some other semi-quantitative measure of the benefits or cost of a resource. He considered the key issue for effective fisheries management is to develop, rather than look at problems now, likely 10-20 year *scenarios* for a particular fishery and begin working with various users in that fishery to think through those scenarios so that strategies can be developed before problems arise.

Russell Reichelt opened the *general discussion* of Session 6 by commenting that multiple use leads to multiple owners, and, in the habitat management scene, multiple regulators eg health, environmental and fisheries agencies, which would indicate that institutional arrangements are a key to effective management. Ross Winstanley was concerned that this approach to the management of aquatic resources, which encompasses a number of users besides just fisheries managers, requires an institutional framework where either single bodies or networks of agencies are organised and able to work cooperatively. This would seem to indicate that the Australia's current system is heading in the wrong direction. Andrew Staniford pointed out that this is only one approach of three he presented and which may not necessarily be recommended, but we need to look at what options are available to make our current institutions more effective.

Jeremy Prince felt that in contrast to the protected area approach in which nobody owns or cares for the resource, the value-added approach is superior as it is in the owner's long term interest to care for and monitor the resource.

John Glaister said that in industries where user group interaction was well developed, such as the US Forestry, user group participation generated more conflict with the industry managers than before user groups became involved. He repeated earlier comments that fishery management is getting away from the role of fisheries biologists but not fisheries scientists.

Jim Puckridge believed that multiple use has the potential for education and resource ownership but he expressed some caution that long term ownership in agricultural industries has not necessarily resulted in the conservation of the resource. Russell Reichelt commented that, in his understanding, the multiple use system on the Great Barrier Reef was introduced to minimise conflict rather than underpinning any fundamental set of strategic principles other than to manage the reef wisely.

Barbara Richardson suggested there is scope in both the multiple use protected areas and the value-added approach. What is required is to come together in a planning process as has been done in terrestrial systems, to consider multiple use zoning that takes into account the different sensitivities, the appropriate uses of those areas, and the conservation and sustainable use of those areas. The planning framework on an institutional basis should have an integrated approach with fisheries management being the major focus. The plan should incorporate other factors including value-added principles in order to make more integrated decisions in all of our waterways. Murray MacDonald stated that the multiple use concept has been discussed at a number of national forums over the last 10 years. These forums have emphasised the need for a strategic planning process to accommodate, besides fisheries, a large range of legitimate and possibly conflicting uses of resources. This planning process should involve participation by both government and community stakeholders. He felt that this process should be emphasised in a number of current initiatives at a national level such as Ocean Rescue 2000 and the National Conservation Strategy.

Bryan Pierce asked Andrew Staniford to cite examples from other natural resources where the value-added approach has worked to enhance the sustainability of that resource. Andrew Staniford replied that the developments in the tuna fishery since quotas were introduced has given it a greater chance of being sustainable. The emphasis is now placed on cultured fish and how they complement the wild stocks. He also agreed with the comments by Murray MacDonald on the advantages of the strategic planning process and stated that unless all stakeholders are involved, the plans will be developed by politicians with the interest of political lobby groups tending to dominate.

Jeremy Prince commented that if we are able to put value to resources, the strategic planning process would become bogged down in community and government committees. Rather a group of individuals should actually own the resources thereby attaching their own value and ensuring the sustainability of the resource. Murray MacDonald said that Jeremy Prince's comments were fine if you make the assumption that fisheries is the only use of those resources and the community who own the common property resource are quite happy to relinquish their ownership to hand it over to a particular user group. This is not always the case. He went on to say that the value-added approach can be successfully integrated into the strategic planning process if it is defined, rather than in economic terms, as what the various stakeholder's interests are and the perceived benefits they receive from the various uses of the resource.

Russell Reichelt asked Peter Young how much strategies such as Ocean 2000, Coastal Zone Strategy influence CSIRO's research planning on environmental issues. Peter Young stated that he wished CSIRO was more involved in these initiatives. He also emphasised some caution over property rights as it may be economically viable to fish a stock down, then get out as was practised on whales along the east coast of Australia.

David Pollard supported previous comments that a much broader value-added approach, rather than a straight economic approach, needs to be taken as you may preclude other alternative uses for a short term economic gain, eg. fishing *vs* fish-watching. Andrew Staniford distinguished between commercially driven values and what he referred to in his talk as a value-added approach that includes non-commercial values. He agreed with Jeremy Prince that property rights can be effective in some situations but in others there is a need to develop a common property approach where the users of the resource have shared ownership of the problems and solutions and are committed to implementing these.

Russell Reichelt asked Rob Lewis to what extent would the development of the National Coastal Strategy or Ecologically Sustainable Development principles assist his department with the development or loss of sea grass areas in South Australia. Rob Lewis thought these strategies have been very valuable to him as a planner, particularly Ocean 2000 in that it deals with issues and involves the participation of commonwealth, state and local user groups. This has been the approach of the marine scale fishery in South Australia and the industry has been fully supportive of this method.

Jeremy Prince explained that fishing down a resource for short term economic gain is valid if you have a lot of virgin biomass but such situations are rare in the current day. It is of long term benefit to buy up undervalued resources and rehabilitate these, whereas the common situation discourages people receiving the benefit of their own rehabilitation efforts. David Pollard refuted Jeremy Prince's comments by suggesting that a broader approach than resource ownership is required and no one should own a resource because it precludes all other possible alternative uses in perpetuity. Andrew Staniford agreed with these comments.

Russell Reichelt stated that at present the fishery scientist is held back because of the lack of long term data sets on which to base decisions. He asked Andrew Staniford who are the people best suited to adopt the long term view on issues such as data collection and resource management. Andrew Staniford replied by saying that because of the common property value of fish resources there aren't any incentives to keep long term data sets so there's a strong basis for government intervention in the collection and distribution of these data. Diane Hughes said there is a strong case for education of the users of a resource and the public on how to recognise and improve the value of an asset.

Murray MacDonald asked David Pollard to comment on how effective had a strategic management framework been in achieving equity in the allocation of users in the Solitary Islands marine protected area and to what extent had the value of various habitats and resources in that area been recognised. David Pollard considered that the Solitary Islands marine protected area, through public participation in the multiple use zoning scheme, is quite good in its general aims at sustainability and multiple use equity. He was concerned however that the area may not sustain long term sustainability in biodiversity as it did not contain large protected areas.

Barbara Richardson supported David Pollard's comments in saying that the Solitary Islands marine protected area has been a valuable process not just in terms of the interaction between user groups, identifying management objectives and ownership but also for fisheries managers in receiving a cross fertilisation of ideas and feedback in ways to best manage or improve management in the future.

SESSION 7

MANAGEMENT (AMELIORATION, ENHANCEMENT, CONSERVATION)

Session Chairperson: B.A. Richardson
Session Panellists: R.N. O'Boyle
J. Burchmore
W. Fulton
Rapporteur: D.A. Pollard

CHAIRPERSON'S INTRODUCTION

B.A. Richardson
NSW Fisheries
Locked Bag 9
Pyrmont NSW 2009

Introduction

The management of fisheries cannot be effective without the integration of stock management with management of the habitat and environment upon which these living systems depend.

As information grows, the relationships between fish populations and fish habitats become better understood but there remain many questions yet to be answered. In the meantime policies and management programs need to be implemented to protect and conserve, and where appropriate restore, remaining fish habitat quality and quantity.

In developing policies the objectives must be clearly defined and the expected outcomes identifiable or measurable. Fish habitat management programs need to set a policy framework including the general underlying question, i.e. *What is happening to fish habitats and what do we need to achieve?*

Should we aim to:

- restore or rebuild lost habitat?
- maintain the status quo—no further loss?
- maintain the area of critical habitats as understood?
- protect the diversity of habitat types and their relative proportions?

And:

- What environmental factors are central to effective habitat management?

The answers to these questions will depend largely upon what we know now and this is certainly an incomplete understanding.

Canada's Department of Fisheries and Oceans has a policy to increase the productive capacity of habitats for the nation's fisheries resources. This has as its guiding principle that there be no net loss of fish habitat and provides for compensatory habitat where losses are likely to be incurred through development.

NSW Fisheries has as its major objective to protect the diversity of habitats as functional units with special emphasis on protection and restoration of critical habitats.

Habitat management strategies

Strategies involved in habitat management programs include amelioration, enhancement and conservation measures. For the purposes of this workshop, amelioration is defined as measures to improve a degraded environment or restore a habitat. Enhancement can be considered as the means to add some additional productivity. Conservation is the suite of measures that aim to protect the remaining natural habitats and ecosystem integrity.

Determining when and where these strategies are to be used requires localised knowledge, accurate recognition of the problem and a capacity to monitor the ecosystem response.

Amelioration

The types of habitat degradation or inadequate environmental management which are appropriate for amelioration include:

- controlling point source pollutants—pollution reduction programs;
- restore flushing and tidal exchange to impacted wetlands or landlocked coastal lakes;
- revegetation of aquatic and riparian habitats;
- return streamflows to meet instream needs;
- provision of fish passage facilities.

Who's responsible?

The legislative provisions and agency responsibility may well lie in several different government portfolios. Furthermore these programs may also require a coordinated effort between private companies, individual land owners, and local government. Often a combination of policies and legislation is required to achieve amelioration.

Who should pay for amelioration?

The costs of undertaking these programs could be subsidised by levies on polluters, poor land managers, large scale water users, etc.. Given the problem of identifying individuals responsible in the majority of cases, these costs are more commonly borne by governments. This may partially account for the limited amelioration which is undertaken.

Enhancement

Enhancement is usually applied where a particular opportunity exists or is created. Compensatory habitats, artificial substrates and man-made habitats, or stocking programs could be included in this category. These opportunities are a valuable learning tool as well as a step towards returning some component of lost productivity.

The objective in this strategy would be to increase the ecosystem productivity, fish abundance or diversity, or possibly to replace or increase the area of a particular habitat.

The scope to undertake enhancement may be affected by whether or not the owner of the land, lake, seabed etc. is in agreement. Generally most works would be undertaken on Crown-owned lands or leaseholds. This is the case for a project being undertaken on Kooragang Island in the Hunter River, NSW, whereby leasehold land is proposed to be re-levelled and channels created to provide for wetland creation.

Compensatory habitat creation is a tool which could be legislated for as a means to amend damage from illegal activities or restitution. The more likely application is the negotiation of compensatory habitat with developers who seek to destroy or modify an area of naturally occurring habitat. The costs are met by the proponent developer in a form similar to "polluter pays". One needs to ask: Is this likely to be as effective as natural habitat, and if not, how much replacement habitat would be appropriate?

Conservation

One of the first questions to consider is: Should we concentrate on species conservation or ecosystem conservation? The capability of the aquatic environments to be productive and sustain fisheries is dependent on a functional ecosystem. However, in the case of endangered or threatened species, it is difficult to manage the problem solely by general ecosystem conservation.

A strategy for conservation of habitats and ecosystems is needed in day to day management programs. Without complete knowledge of the ecosystem, I believe it is most important to maintain the diversity of habitats and the proportional representation wherever possible. This begs the question: Do we know enough to only protect perceived critical habitats?

Conservation can be undertaken in the form of guidelines and policies for assessing developments and environmental impacts. The objective in this strategy is to identify any activities which may have unacceptable impacts on aquatic ecosystems or to minimise the impacts arising from developments which are seen as justification in the public good. Fisheries legislation and environmental planning legislation are both important to achieve this. However this alone could continually reduce the quality and quantity of fish habitats without the balance of amelioration and enhancement strategies.

The development of a representative reserve system which would allow long term conservation of habitats and communities is one appropriate conservation measure which can be an effective addition to fisheries closures which protect habitat. The size and selection of reserves can vary but there seem to be greater benefits to be derived from larger areas with some form of buffer area than small isolated pockets as protected areas.

A further means of protecting fish habitats can be by environmental protection zones incorporated in local and regional planning instruments as well as reserves and closures under fisheries legislation. Community education and awareness is most important in delivering conservation programs to greatest effect (Moberly this meeting). The public can participate in the planning process and raise the level of performance of local governments in protecting fish habitats. Furthermore, the community, including user groups such as fishermen or passive users such as conservationists, divers etc., can participate in habitat management programs, eg. planting mangroves, clean up programs.

The above ideas are suggested strategies and concepts and are by no means comprehensive. The workshop session may challenge and debate these issues with a view to proposing more effective means of sustaining fish habitats.

THE MANAGEMENT OF MARINE HABITAT

R.N. O'Boyle

Marine Fish Division
Bedford Institute of Oceanography
PO Box 1006
Dartmouth Novia Scotia Canada B2Y 4A2

Abstract

The management of marine habitat involves issues that are chemical, biological and physical in nature. The variety of impacts and the complexity of marine ecosystems presents special challenges to habitat managers and calls for careful attention to management functions and institutions. Canada adopted a new approach to habitat management in 1986, elements of which are discussed in this paper. The main features of this policy relate to the objective of overall net gain in productive capacity and how this is achieved through conservation (no net loss principle), restoration and development. Models that quantify these principles and thus provide targets for management are only now appearing, examples of which are given for aquaculture sites. These represent a good starting point for the development of more comprehensive views of the ecosystem both in relation to aquaculture and elsewhere.

An extensive set of regulatory measures has been established to provide managers with the tools to manage habitat development towards the objectives. However, in the case of the offshore marine environment particularly, much needs to be done to ensure the long-term monitoring of human impacts on the habitat. Recent initiatives in this are presented. Current efforts to integrate the disparate data bases used in support of habitat decision-making are also discussed. The paper ends with a description of the consultative process used by Canada to review proposals that have potential habitat impacts.

Introduction

In Canada, as in other parts of the world, marine habitat problems can be considered under three broad categories. Chemical issues involve the impacts of oil spills, both from rigs and vessels, and exploration. The ocean dumping of chemical waste can be considered under this category. The second set of issues are biological in nature, generally involving the effects of eutrophication on coastal waters, resulting from runoff of agricultural land, sewage from urban areas and industrial waste from fish processing, fertiliser plants and even aquaculture. For instance, there has been much discussion in Europe on the linkage between human-generated eutrophication and phenomena such as red tide and the incidence of Phocine Distemper Virus (Ross *et al.* 1992).

Finally, there are the impacts of physical structures on the habitat. These include barriers (Canso Causeway, link to Prince Edward Island, dam for Fundy Tidal Power), marine debris (plastics, discarded nets which can "ghost-fish") and disruption of the benthos (ocean mining, fishing gear impacts on bottom, laying of cables and pipelines, etc). A number of these are detailed in Messieh *et al.* (1991).

All these issues are common to most parts of the industrialised world and have elicited various management activities as and when they arise. However, to be truly effective, the response of managers to habitat issues must be based on a system which has well defined goals, regulations and monitoring activity. Otherwise one runs the risk of a patchwork approach to habitat management.

In 1986, Canada adopted a new approach to habitat management, after an extensive period of consultation. In this paper, I will discuss the main elements of the current policy and its implementation. As well, problem areas will be identified together with some of the current programs that are underway to address these.

The management unit

Any discussion on the management of habitat—marine or otherwise—must start with a definition of habitat. In the Canadian Fisheries Act, habitat is defined as "spawning grounds and nursery, rearing, food supply and migration areas on which fish depend directly or indirectly in order to carry out their life processes". Thus, it is that part of the ecosystem, including the biotic and abiotic components, upon which a fish depends. The Act is silent on the issue of the importance, commercial or recreational, of the fisheries and indeed is broad enough to include the entire ecosystem. In recent years however, the management of fish habitat has been directed towards those species that are economically or socially important to Canadians. It is becoming increasingly evident that many user groups are interested in access to the marine environment, most notably tourist interests, and there is growing concern that degradation of marine habitat will, in the long term, have negative impacts on other parts of the environment. It is evident then that as the use of habitat evolves, so too will the focus of its management. Current Canadian legislation appears flexible enough to accommodate this. In addition, as with fisheries management, it is important to keep in perspective the economic and social consequences of habitat degradation, and thus its management, on the human population.

Given that habitat management can be considered synonymous with ecosystem management, it is easy to understand why it has been so very difficult to undertake. In fisheries, the focus of management is the biological stock. This considerably simplifies the situation in relation to goal setting, regulation and monitoring. With habitat, on the other hand, even defining the ecosystem is not a trivial matter. Nevertheless, it is axiomatic that in order to effectively manage, the first step is to define the boundaries of the "management unit". Too often, this has been given little consideration in the definition of habitat management systems.

Off Nova Scotia, progress has been made on the preliminary identification of a number of ecosystems (Mahon and Sandeman 1985). The Gulf of Maine ecosystem, characterised by a mixture of warm and cold water species, stretches from Cape Cod to midway up the Scotian Shelf (Figure 1). The Laurentian Ecosystem continues from there, across the Laurentian Channel along the southern coast of Newfoundland to meet the Labrador Ecosystem which continues north up the Labrador Shelf. The boundaries of these ecosystems tend to coincide with major physical oceanographic features present on the coast, lending support to the hypothesis that the productivity of these communities is largely related to the prevalent environmental conditions.

As yet, those ecosystem boundaries have not been used to assist in the management of offshore habitats as more work is required to confirm and refine the integrity of these units (Mahon et al. 1984). Nevertheless, habitat managers will have to be aware of the characteristics of the ecosystem with which they are dealing. Definitions of characteristics, such as the ones presented, are a useful first step in this regard.

The goals of habitat management

Management goals can be considered as being composed of objectives which define what it is that one wants to achieve, and strategies which are quantifiable targets or constraints by which one measures the success (or otherwise) of management.

In Canada, the long-term objective is the achievement of an overall net gain in the productive capacity of fish habitats (Anon. 1986). This is considered to be possible for anadromous and certain freshwater and shellfish species but more limited in marine ecosystems. Nevertheless, the long-term thrust of the policy is to regain habitat to compensate that which has already been lost.

This objective is achieved through three initiatives:

1. **Conservation**—Maintain the current productive capacity of fish habitats supporting Canada's Fisheries resources, so that fish suitable for human consumption may be produced. This strategy is synonymous with the No Net Loss Principle.

2. **Restoration**—Rehabilitate the productive capacity of fish habitats in selected areas where economic or social benefits can be achieved through the fisheries resources. This is to correct past mistakes (ie installation of dams without fish passages, sewage treatment plants, etc).

3. **Development**—Create and improve and generally enhance fish habitats in selected areas where the production of fisheries resources can be increased for the social or economic benefit of Canadians.

The combination of conservation, restoration and development initiatives leads to a net gain in overall habitat productivity. While this is simple in concept, it does not help unless we have targets and constraints by which we can define and measure levels of conservation, restoration and development.

In fisheries management, mathematical formulations called strategic models, such as Surplus Production and Dynamic Pool models (O'Boyle 1987), have been used to define targets and constraints on harvesting such as Maximum Sustainable Yield, F_{MAX} and minimum Spawning Stock Biomass.

These models summarise our knowledge of the dynamics of marine populations and provide understanding on the impacts of harvesting on yield, recruitment, population growth and abundance and so on. They thus provide the means to quantify overfishing, and therefore provide guidance for the control of harvesting levels.

Development of strategic models with the ensuing targets/constraints for habitat management, is not as advanced as for fisheries management. There is an urgent need for these, given the rapidly growing encroachment on habitat that is occurring around the world. Given the complexities of the ecosystem it is understandable that model development has been slow. This is not to say that there has not been progress. Gordon (1992) for instance, reviews a number of significant efforts in the North Atlantic. Much more is still required. There may be a temptation to resist model development until more data are collected. This would be a mistake. Silvert (in press) shows that it is important to create models, based on first principles and simple assumptions now, that can be updated as more is learned about the system in question. Through the constant interaction between field and model, understanding of the cause and effect of habitat management can be dramatically improved.

Silvert (in press) illustrates how simplified models can be constructed to evaluate the impacts of salmon aquaculture sites on habitat. As with fisheries models, the impacts can be evaluated at various levels. This is an important consideration in that it partitions the overall problem into smaller, more manageable units and is perhaps an approach that could be adopted for other habitat problems.

The first level that he considered was at the inlet or regional scale. He described total acceptable bay production as a function of depth, flushing time, nitrogen production and the capacity of the bay to handle this production, or

$$\text{Maximum Production per Unit Area} = f\left(\frac{\text{Depth}}{\text{Flushing Time}}\right)\left(\frac{\text{Nitrogen Load Limit}}{\text{Nitrogen Production}}\right)$$

All these parameters are measurable or can be derived from previous studies. The maximum aquaculture production allowed in the bay that maintains *conservation* of the habitat can then be calculated.

The next level considered was at the local scale involving inter-site impacts and is therefore concerned with placement of aquaculture sites within the inlet. At this scale the localised effects of benthic disposition are important, which can be mitigated against by defining a minimum site depth for a particular size and level of site productivity. Silvert (in press) describes the minimum allowable site depth as a complex function of total production, settling speed, aquatic transport and bottom type. Again, the manager can then judge the appropriateness of a particular cage site at a given depth.

The last scale to consider is at the internal or cage level where oxygen depletion within the cage site itself must be considered. Here, the size of the farm is related to its production capability, or

$$\frac{\text{Maximum Production}}{\text{Length of Site}} = f\left(\frac{\text{Current Speed} \times \text{Depth}}{\text{Oxygen Consumption}}\right) \times \left(\text{Ambient Oxygen Level} - \text{Lowest Acceptable Oxygen Level}\right)$$

Thus, the physical dimensions of the farm can be modified to be compatible with the existing environmental conditions.

At present, while there is recognition that these levels interact, little has been done to quantify this. This is an important area of study. There may be a tendency in the evaluation of small, localised development proposals to grant approval where effects are too small to measure. Over the long term, the cumulative effects of this process will be felt. Thus there is real benefit to providing a means of measuring and assessing this incrementalism.

The above models were formulated for aquaculture applications and are at an early stage of development. There is a need to develop models for other areas of habitat usage so that managers can be provided with quantitative measures of conservation, restoration and development.

The controls of habitat management

There are two aspects of control—regulation to ensure that habitat developments are consistent with the goals of management, and monitoring to ensure that the regulations are having the desired effect. The monitoring activity may lead not only to changes in the regulations but also, through scientific research, to modifications in the strategic models.

Regulation

When a development proposal is received by the Canadian Department of Fisheries and Oceans (DFO), a hierarchy of regulatory measures is used to ensure attainment of the conservation (no net loss) objective:

1. The natural productive capacity must be maintained through redesign of the proposal, location of the project at alternate sites and, in the case of chemicals, pollution control devices. If this is not possible (for other than liquid waste discharges) then,

2. Compensatory options are pursued, including like-for-like habitat on the same site, like-for-like habitat at an alternate site or finally enhancement of the productivity of the existing habitat. If these options are not possible; then,

3. Artificial production may be considered as long as the objectives of local fisheries are observed and the methodology is practical and proven.

There is a considerable package of legislation (Anon. 1991) which provides the enforcement tools of regulation including fines, the requirements for safe fish passage, minimum flow requirements, fish-way protection, physical disruption of the habitat and its pollution. Most of these regulations are more relevant to the inland and coastal waterways but can be extended to the marine environment as required.

There are a number of reporting requirements in the Canadian Habitat legislation that are of importance. Besides the ongoing requirements, it is the responsibility of the developer to provide the analyses of potential impacts of the development on the habitat. This will be discussed further below.

Monitoring

For the marine habitat, most of the monitoring effort has been conducted close inshore. For instance, there is an extensive network of monitoring sites to evaluate levels of coastal phytoplankton, with the focus on those capable of producing phytotoxins. Over the long term, this information will be useful in evaluating whether or not these events are related to anthropomorphic trends. For instance, it has been proposed that the mass mortality of harbour seals in Europe in 1988, caused by Phocine Distemper Virus, might be related to the levels of immunosuppressive pollutants (Ross *et al.* 1992).

In the offshore area, while effects of bottom disturbance have been documented (Messieh *et al.* 1991), there has been no long-term program in place to track decadal scale changes. However, as a consequence of consultations conducted in 1989 (Hache 1989), a program has been established to evaluate the impact of trawling on the benthic habitat. Thus far, the project has focused on the development of gear capable of sampling the bottom, including side-scan sonar mounted on a towed vehicle, epibenthic sled equipped with video camera and hydraulic-powered, video-equipped bottom grab (O'Boyle *et al.* in press). This project will use the opportunity provided by areas closed to fishing activity to investigate the impacts of fish trawling on the benthos, similar to the studies conducted by Sainsbury (1991). In the longer term, it may be possible to integrate this benthic monitoring activity with annual groundfish trawl surveys.

An important aspect of monitoring is, as stated earlier, the provision of a decision-making facility to managers to assist in the initial review of development proposals and their ensuing impacts. In fisheries, extensive information systems have been developed to provide the necessary input to management. Comparable systems are only now being developed for habitat applications.

One of these is being developed as part of a project, on the L'Etang estuary in southwest New Brunswick, to provide the basis of the evaluation of the suitability of aquaculture sites. The work of Silvert (in press) and Silvert *et al.* (1990), mentioned earlier, is part of this project.

The structure of the decision-making system currently under development is presented in Figure 2 (Keizer, pers. comm.). The central feature of the system is a GIS-based analysis package. This package can access a number of data bases, ranging from human activity to fisheries statistics, available in a number of different organisations and locations, and allow the collation of disparate data over different spatial scales. Interestingly, this requirement, driven by the complexity of habitat issues, has called for the definition of computer communication standards, policies on access to data bases, and consideration of institutional arrangements to facilitate data sharing.

On the other side of the GIS are decision-making models which would be unique for each habitat issue. For aquaculture, these models would allow the manager to survey all information relevant to a specific locale under consideration for site placement, as well as accessing other models available to evaluate impacts. Initially, the system may only be a pointer to these models but in the future it will allow options investigation by the decision maker.

It is envisaged that this information system will become a key component of the evaluation process for habitat proposals. Nonetheless, while the system will provide the necessary technical backup, an extensive consultative process is also required to evaluate habitat-related proposals. Experience has shown with other management systems that consultation is a key element that requires careful consideration (O'Boyle submitted).

The consultative process

The process used by Canada to evaluate development proposals is given in Figure 3. All Canadian federal government departments are required to conduct evaluations for projects under their jurisdiction as part of what is referred to as the Environmental Assessment Review Process (EARP). The initial information and evaluation report (Environmental Impact Statement) prepared by the project proponent is received by the DFO Habitat Management Branch for their evaluation. A considerable consultation process now commences during which the potential biological, chemical and physical impacts of the project are assessed. It is during this phase that the objectives of conservation, restoration and development are assessed. Until recently, regional scientists had to undertake these evaluations. In 1992, the Canadian Atlantic Fisheries Scientific Advisory Committee (CAFSAC) struck a new subcommittee on habitat to serve as a scientific peer review forum on habitat issues. This pooling of scientific expertise to consider resource issues has been central to the success of other management agencies such as North Atlantic Fisheries Organisation (NAFO), International Council for the Exploration of the Sea (ICES), Canadian Atlantic Fisheries Scientific Advisory Committee (CAFSAC) and International Commission for the Conservation of Atlantic Tunas (ICCAT), and will greatly facilitate efforts to provide informed advice on habitat management.

This process can either lead to further public consultations (major impacts projected) or more limited consultation with the immediate proponents (minor impacts projected). If public consultation is required, the Minister of Fisheries and Oceans recommends to the Minister responsible for the Federal Environmental Assessment Review Office (FEARO) that a public review panel be appointed. The latter is a non-partisan board of experts to which DFO and the proponents, as well as the general public, make submissions. This board then considers the input and makes a recommendation to the Minister. The proponent can appeal this decision directly to the Minister. Ultimately the latter makes the final, binding decision.

While this process is involved, it has worked relatively effectively in decision-making on habitat issues.

Concluding remarks

The management of Marine Habitat is complicated by the scale of the problem and complexity of ecosystems. However, this enforces the need for a structured approach to management involving consideration of the management units, objectives, targets, regulations, monitoring and consultative processes. Canada has made significant progress in establishing a management policy, regulatory framework and consultative process to guide its management of habitat. Considerable work is still required to define models of the cause and effect of impacts as well as to establish effective monitoring systems. Work in these areas has been initiated and will require a sustained effort to ensure the realisation of an effective system for marine habitat management.

Acknowledgements

The author gratefully appreciates the input and review comments received from Don Gordon, Bill Silvert, Paul Keizer and Dave Marantz.

References

Anon. (1986). The Department of Fisheries and Oceans Policy for the Management of Fish Habitat. DFO/3524 Minister of Supply and Services Canada. 1986. Cat. No. Fs 23-98/1986 E.

Anon. (1991). Canada's Fish Habitat Laws. DFO/4438. Minister of Supply and Services Canada. 1991. Cat. No. Fs 23-40/1991 E.

Gordon, D.C. (1992). Current Applications and Future Developments in the use of Mathematical Models in the Implementation of Monitoring Programs and Regional Assessments. Int. Council for the Exploration of the Sea C.M. 1992/Poll.: 9, Sess. V, Annex 3.

Hache, J.-E. (1989). Report of the Scotia-Fundy Groundfish Task Force. Minister of Supply and Services Canada. Cat. No. Fs 23–157/1989 E.

Mahon, R. and E.J.S. Sandeman (1985). Fish Distributional Patterns on the Continental Shelf and Slope from Cape Hatteras to the Hudson Strait: A Trawl's Eye View, pp.137-152. *In*: R. Mahon [ed] *Towards Inclusion of Fishery Interactions in Management Advice. Can. Tech. Rep. Fish. Aquat. Sci.* No. 1347.

Mahon, R., R.W. Smith, B.B. Bernstein and J.S. Scott (1984). Spatial and Temporal Patterns of Groundfish Distribution on the Scotian Shelf and in the Bay of Fundy, 1970-81. *Can. Tech. Rep. Fish. Aquat. Sci.* No. 1300.

Messieh, S.N., T.W. Rowell, D.L. Peer and P.J. Cranford (1991). The effects of Trawling, Dredging and Ocean Dumping on the Eastern Canadian Continental Shelf Seabed. *Continental Shelf Research* **11**, 8–10.

O'Boyle, R. (1987). Approaches to Fisheries Management in the North-Western Atlantic—A Canadian Perspective. Marine Law Institute, Univ. Maine School of Law, Portland, Maine, 482 pp.

O'Boyle, R. (submitted). Fisheries Management Systems. A Study in Uncertainty.*Can. J. Fish. Aquat. Sci.*

O'Boyle, R., K. Drinkwater, B. Petrie, T. Rowell and P. Vass (in press). The Atlantic Fisheries Adjustment Program. Science Review 1990/91.

Ross, P.S., I.K.G. Visser, H.W.J. Broeders, M.W.G. van de Bildt, W.D. Bowen and A.D.M.E. Osterhaus (1992). Antibodies to Phocine Distemper Virus in Canadian Seals. *Veterinary Record* **130**, 514–516.

Sainsbury, K.J. (1991). Application of an Experimental Approach to Management of a Tropical Multispecies Fishery with Highly Uncertain Dynamics. ICES. Mar. Sci. Symp. **193**, 301–320.

Silvert, W. (in press). Assessing Environmental Impacts of Finfish Aquaculture in Marine Waters. *Aquaculture*.

Silvert, W.L., P.D. Keizer, D.C. Gordon and D. Duplisea (1990). Modelling the Feeding, Growth and Metabolism of Cultured Salmonoids. ICES. C.M. 1990/F.8. Sers.O.

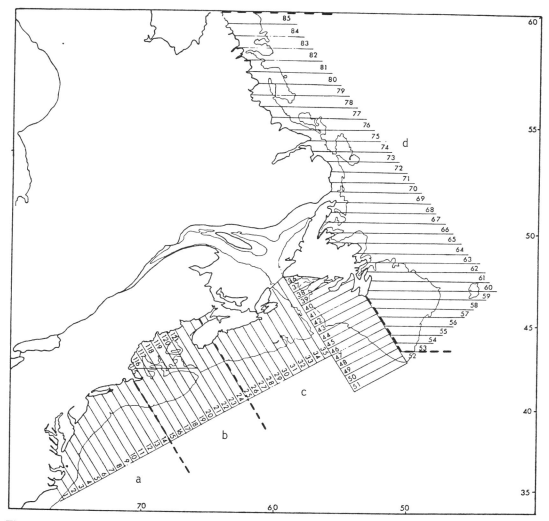

Figure 1. The four major clusters of bands using the occurrence of fishes at depths of 50-200 m to identify the marine ecosystems off the Canadian East Coast; a) New England; b) Gulf of Maine; c) Laurentian; and d) Labrador. (from Mahon and Sandeman 1985).

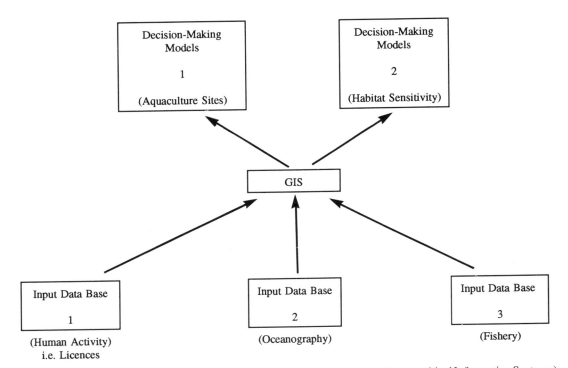

Figure 2. Relationship among input and decision-making data bases and GIS (Geographical Information Systems) developed as a decision-making system for the L'Etang Estuary Project.

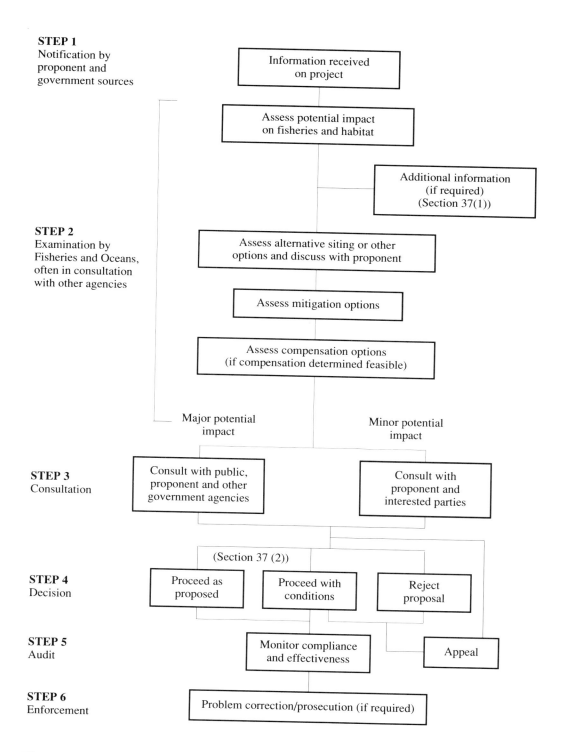

Figure 3. Procedural steps to achieve No Net Loss (from Anon. 1986).

MANAGEMENT OF THE ESTUARINE HABITAT

J. Burchmore
NSW Fisheries
Locked Bag 9
Pyrmont NSW 2009

Estuarine habitats have suffered major disruptions this century, and habitats such as seagrass beds, mangroves and saltmarshes are all continuing to suffer from the cumulative impact of small encroachments—the tyranny of small decisions.

Issues and problems

The importance of estuaries

Estuaries are ecologically significant areas that provide a variety of habitats for aquatic plants and animals. Their importance to fisheries and thus the need for their protection, has only really been realised in the past few decades (e.g. Pollard 1976). These habitats play a major role by providing nursery areas for juvenile fish as well as shelter, food and breeding areas for many adult fish.

Estuaries are areas of considerable significance to both commercial and recreational fisheries. It has recently been estimated that 64% (by value) of the 1987/90 catch of commercially marketed species in New South Wales are dependent on estuaries during all or part of their life cycles. This equates to a value of $77 million (C. Copeland and D. Pollard, pers. comm).

Estuarine habitats

Estuarine fish habitats can be roughly differentiated by vegetation type, sediment type and depth. They include seagrasses, rocky reefs, mangroves, unconsolidated sediments of sand or mud, and saltmarshes. Of these habitat types, the most threatened are probably seagrasses, mangroves and saltmarshes.

Seagrasses in particular have suffered serious declines. Of the major estuaries in NSW which have been intensively studied, many have lost as much as two thirds of their seagrass beds over the last thirty or forty years. For example, in the Clarence River in NSW there were 356 ha of seagrasses in 1942 and only 158 ha in 1981 - a 56% decrease (Shepherd *et al.* 1989). More recent research has shown that by 1990 there had been an 80% decrease on the 1942 figure. Over 90% of the wetlands of the Hunter, Clarence and Macleay River floodplains have been affected by drainage and flood mitigation schemes.

Activities liable to affect estuarine habitats

There are a number of human activities that have the potential to severely impact on estuarine habitats (Pollard *et al.* 1991). The key ones in NSW are:

- infilling, dredging and extractive operations;

- diffuse and point source pollution;

- waterfront developments, such as canal estates;

- marinas and other recreational facilities, such as jetties;

- road and bridge construction; and
- structural flood mitigation works.

Management problems

Conservation, amelioration and enhancement of these important fish habitats are honourable ambitions but there are many hurdles to overcome before these can be achieved:

a) There is either a lack of adequate legislative protection for fisheries habitats, or legislation may rest with an inappropriate agency where this provision is not a high priority. For example, there is no direct legislative protection for seagrasses in NSW.

Neither is there legislation which requires compensation/mitigation for habitat destruction. Instead, there is a reliance on time-consuming negotiations throughout planning processes; e.g. during the public exhibition and determination phase of developments requiring environmental impact statements.

b) The final decisions with regard to the destruction or protection of important fish habitats often rest with other agencies (e.g. construction authorities such as the Department of Public Works) or with different levels of government (such as local councils). Often the government agencies with the largest budgets are in the strongest positions of control and influence.

c) There is a critical lack of good information bases, particularly on mapping of habitats, habitat utilisation, necessary buffer zones, etc. Estuarine inventories are perhaps the most valuable management tools, but these are few and far between and are seldom upgraded. There is also a critical lack of restoration techniques.

d) There is great difficulty in managing cumulative impacts resulting from the "tyranny of small decisions". It is very difficult to argue the case for the retention of *one* mangrove or a small patch of seagrass that may be lost as the result of a development. As a result, over time large areas of important habitat are gradually being eaten away.

e) There is a lack of economic and cost/benefit studies. These can rely on commercial fish catch data but this is usually only an annual estimate. This then ignores the fact that fisheries are renewable resources if managed properly, and therefore have potentially an infinite value to the community. The values of other functions such as stabilisation and filtering are often ignored and not costed out.

f) In mitigation projects there are seldom any acceptable measures of success, e.g. "x% of biodiversity returned after y years". Often there are no baselines to determine appropriate performance indicators and it is often necessary to rely on the results of only very brief surveys carried out in relation to environmental impact assessment.

Management solutions

Today as managers of estuarine fish habitats, we are faced with the fact that we may be lacking necessary background information, legislation, decision-making ability and resources. What we *do* have, however, is the vision and the ideas for habitat conservation, amelioration and enhancement.

Every opportunity must be taken to participate actively in the processes, whether it be attending interdepartmental or catchment management committees, preparing submissions in response to environmental impact statements and other planning instruments, presenting evidence for Commissions of Inquiry, or merely undertaking day-to-day negotiations with other agencies.

There are three current case studies in NSW estuaries which, to varying degrees, all involve concepts of habitat conservation, amelioration and enhancement. These may help to highlight some of the problems and complexities inherent in proactive habitat management.

Case Study 1—Ballina Mangrove Compensation Project

Development: Bridge and access road through mangroves.

Impact: Loss of 8 ha of mangroves.

Management Action: Development approved by local council on the condition that an equivalent area of mangroves be created.

Problems: No available area was available and there were no available techniques or baselines.

Solution: Removal of fill from previously reclaimed industrial land, with various treatments being used to determine best method for future projects, including:

- transplanted trees (1-3m).
- transplanted seedlings (20-50cm).
- planted seeds.
- natural recruitment.

Results: Planted seeds were found to be the best technique, and the resulting mangroves are now growing successfully on the reclaimed areas. A large and diverse fish population has also returned to the area.

Case Study 2—Kooragang Island Wetland Restoration/Compensation Project

Development: Historical reclamation of islands in the Hunter River for industrial land. Wetland areas were also drained for grazing.

Impact: More than 600 ha of mangroves and saltmarsh has been reclaimed or degraded. Tidal creeks have been blocked by poor culvert design.

Management Action: NSW Fisheries approached key players with the concept of repairing past damage and creating new habitats. A jointly-funded feasibility report was prepared.

Problems: Community/landowner acceptance. Cost effectiveness of earthworks. Continued access for utilities.

Solution: The feasibility report recommended opening and deepening of tidal creeks to allow tidal inundation and use of experimental plots to determine the best way of restoring/creating fish habitats.

Results: Unknown at this time.

Case Study 3—Construction of Third Airport Runway in Botany Bay

Development: The third airport runway which will encroach into Botany Bay.

Impact: Large scale dredging (15 million cubic metres) and reclamation. Loss of 30 ha seagrass, 4 km of sandy beaches, destruction of fishing grounds.

Management Action:	Plan of Management requiring monitoring and compensatory mechanisms.
Problems:	Lack of appropriate compensatory mechanisms, especially for large-scale dredging. Little available room to manoeuvre. Lack of performance/success indicators.
Solution:	Equivalent seagrass replacement and possible rocky reef creation. Potential for restoring with fish.
Results:	Unknown at this time.

Conclusions

The challenge is not merely to conserve what is left of our important estuarine fish habitats but, whenever and wherever possible, to repair past damage done and compensate for unavoidable destruction. The opportunities exist now, and fisheries habitat managers need to be entrepreneurial in their approach and to broaden their traditional horizons and skills.

These estuarine habitats are the source of tangible benefits to the public. They, together with the fisheries they sustain, are a renewable resource of potentially infinite value to the community. They are a common property resource and the whole community must share the responsiblity for conserving them.

References

Pollard, D.A. (1976). Estuaries must be protected. *Australian Fisheries* **35(6)**, 1-5.

Pollard, D.A., M.J. Middleton and R.J. Williams (1991). Estuarine Habitat Management Guidelines. NSW Agriculture and Fisheries, Sydney.

Shepherd, S.A., A.J.McComb, D.A. Bulthuis, V. Neverauskas, D.A. Steffensen and R West (1989). Chapter 12 Decline of Seagrasses. In: *Biology of Seagrasses* Ed A.W.D. Larkum, AJ McComb, and S.A. Shepherd. Elsevier, Amsterdam.

MANAGING FRESHWATER FISH HABITAT

W. Fulton

Inland Fisheries Commission
127 Davey Street
Hobart TAS 7000

Issues/Problems

There are many factors that may influence freshwater fish habitat quantity and/or quality, including:

- water abstraction:
  - agriculture;
  - industry;
  - power generation;
- impacts on habitat—(direct or indirect):
  - mining;
  - industry;
  - forestry—chemicals, physical changes;
  - agriculture—clearing, chemicals;
  - power generation—temperature, discharge;
  - government—waste disposal;
  - catchment management;
  - stream 'improvement';
  - fauna translocations.

This list can essentially be viewed as a list of groups to deal with in user-conflict situations. Nowadays one should also certainly add conservation groups.

In trying to look after fish habitat interests in these conflicts, managers are immediately confronted by lack of information.

What exactly is the problem?

- Is the fish habitat being affected directly or indirectly?
- Is the effect lethal or sub-lethal?
- Is the process reversible in the short or long term?

The ability to positively influence fish populations via their habitat requires knowledge of the relationships between a fish species and its habitat and the effects of outside influences on that habitat.

Decisions are invariably made in the absence of sufficient information. For example what is the shortage of a particular fish species really due to? Is it overfishing, habitat degradation, a combination of both, or even over-expectation? The latter is a common problem in recreational fisheries.

The present situation is that regular conflicts are arising between users of freshwater habitat, and resolutions are required immediately. Unfortunately there is always going to be insufficient information on fish habitat requirements. It is nevertheless a distinct disadvantage in the negotiating process that we are usually unable to categorically define the link between fish and habitat and rarely able to define links quantitatively.

Environmental flow assessment methodologies provide some hope for informed input to the habitat allocation process but further refinement is required. In some Australian States there is still no legal requirement or basis for consideration of an environmental allocation in any case.

Management options

Amelioration
There are certain categories of actions that could best be described as amelioration. These are usually designed to limit certain influences following laboratory and field evaluation of particular processes. Guidelines or management plans should be developed, preferably with the involvement of those responsible for the problems.

Examples in freshwater in Tasmania:

- pesticide use in agriculture and forestry—develop guidelines for safe use;
- toxicity studies on paper mill effluents—provide feedback to treatment plant design;
- instream flow evaluation—develop environmental flow parameters;
- effects of forestry operations on freshwater—provide feedback to Forest Practices Code;
- storage discharge problems—develop operational guidelines for power stations;
- lakewater level fluctuations—evaluate conditions to achieve balance.

Enhancement
Can the present habitat or fish numbers themselves be supplemented in some way? Ideally this requires detailed knowledge of the relationship between a fish species and its habitat. If this is known then how can this knowledge be used? Processes usually involve manipulation of a particular part of the habitat.

Examples in freshwater in Tasmania:

- spawning habitat enhancement or control for trout;
- provision of artificial substrates for whitebait spawning;
- construction of artificial stream barriers to prevent migration of predatory fish to conserve particular species;
- controlled stocking for recreational species;
- fish passes to allow access to additional habitat;
- water level management to promote macrophyte growth.

Conservation
Would a reserve actually help a certain species? This raises a series of questions as well as actions.

Again, knowledge of the relationship between the species and its habitat is essential. Should we be looking at single species or at general habitat conservation? What are we trying to conserve?

There may be a need to conserve particular elements of habitat for certain life history stages of a species:

- spawning habitat:
  - estuarine marshes for whitebait;
  - instream snags for river blackfish;
- nursery habitat:
  - lake, river and estuarine marshes for some species;
  - marginal vegetation in lakes;
- adult habitat.

Obviously there is a lower limit to the area of habitat that must be conserved for various reasons. For example, in terms of security a small reserve may only serve to focus attention.

In terms of effectiveness is habitat conservation the answer?

- for a migratory species;
- for an overexploited species.

Detailed knowledge of the life history of the species is therefore essential.

In conclusion, there are no general solutions. There are no universal critical factors; they vary with species, with time and with area. Each individual case will require a specific course of action which must include a monitoring program.

The level of resources available will not influence the best option but, in practice they certainly influence the effectiveness of any action.

Education

Community consultation and education is the most powerful means of effecting change although it can take time. It also requires considerable effort on behalf of managers to ensure that these groups are well supplied with information. It is also the best way of influencing politicians.

A good recent example is the Landcare program through which many community groups have obtained funding to rehabilitate certain lands. In Tasmania this has been widely used to fund willow removal and re-vegetation of stream banks.

This area is the most powerful means of bringing about the climate for habitat protection. However, it requires good information generated through specific applied research.

DISCUSSION OF SESSION 7

Recorded by D.A. Pollard

Fisheries Research Institute
PO Box 21
Cronulla NSW 2230

Questions were first addressed to individual panellists, and then followed by more general discussion.

Following *Bob O'Boyle's* panel presentation, Chairperson Barbara Richardson asked for the discussion to consider pollution property rights, and what impacts these may have on policies and management options for fish habitats.

Murray MacDonald opened the discussion by commenting that the legislation referred to by Bob O'Boyle, and on which it was presumed that all these strategies for management were based, was in fact Fisheries legislation and that the definition of habitat used was based on the value of that habitat to fisheries production. He asked whether there is any other legislation in Canada which defines the value of habitat and the management of habitat in terms other than fisheries, and how it interacts with the Fisheries legislation.

Bob O'Boyle felt that he was not close enough to the issue to make an authoritative comment. He knew that the Department of Environment had a lot of air pollution control regulations, but did not know whether all habitat legislation was specifically under the Fisheries Act.

Peter Young asked where industries and other stakeholders would get involved in the process of decision making.

Bob O'Boyle replied that the proponent comes with a package which includes his proposal, and what he feels he has to do under the legislation. It comes to the notice of Fisheries, and basically within the first step the Government examines it, and if it appears that there will be a major impact, then it is opened to a public hearing process. In short, only the proponent and Government are involved until a decision is made as to whether there will be a major or a minor impact. Clearly, with 2000 proposals coming in a year, that could mean a lot of public hearings. So, what are really being concentrated on are the major impacts, and if the Minister for Fisheries says it is a major impact, he establishes a three to five or seven, member panel. Generally it is at that level that you find a good cross section of the people that would be involved in the issue.

Barbara Richardson, in inviting some questions on *Jenny Burchmore's* panel presentation, believed that the speaker had demonstrated very well the position Fisheries Habitat Managers are put in quite frequently in terms of serious decision making and negotiating.

The sorts of information they may have are often insufficient, but nevertheless we've got to move forward; we have to take opportunities that come up and learn from them so that we progressively develop and improve policies for protecting and managing our habitats.

Julian Pepperell told Jenny Burchmore that he was surprised with the lack of legislation to protect seagrasses and mangroves. This is all the more amazing considering that NSW Fisheries can gazette overnight other restrictive regulations on the catching of fish, like bag limits, closed seasons and so on.

Jenny Burchmore responded that there were other agencies which did not believe that regulation to protect vegetated habitats was a Fisheries role. NSW Fisheries is now getting new legislation and hopefully this will provide proper protection for some of these areas. The really difficult thing in getting new legislation designed to protect fish habitat is defining what it is you want to protect, what constitutes critical habitat, and so on.

Peter Gehrke referred to Jenny Burchmore's mention of the lack of guidelines for the creation of compensatory wetlands in Australia. The US Army Corps of Engineers has guidelines to compensate for habitat damage; for example, when extracting soil from one site for major construction projects elsewhere. A comparison of fish nurseries found that habitats created according to Army specifications consistently provided good recruitment to local fisheries. Despite having limitations in Australia, the same guidelines could be used to develop our own requirements for compensatory wetlands.

Barbara Richardson agreed that this was the sort of approach needed now. We have to go out and do *something* based on our present knowledge. We may make mistakes along the way, but if we can build a reasonable experimental design, it will give us directions for the future, or information on how to do it better in the future.

John Koehn commented on *Wayne Fulton's* presentation that the outline presented for Tasmania was basically how things operate as far as priorities go in Victoria as well. You end up following up projects or files or whatever comes up. How do we get around that, given the lack of resources, and try to set our own agendas to get some of the priority things done? He was sure we know the key factors in freshwaters, anyway, and we know the key threats. You can go down the list—1, 2, 3, 4; dams, riparian vegetation, catchments, toxic impacts, whatever they are. How do we break this cycle every time they build a new dam? Maybe next time in Victoria, which may be 20 years away, we might get them to put in a fish ladder and a multi-level offtake; but in the meantime there are 20 dams out there that are still causing havoc. How do we actually break that cycle?

Wayne Fulton believed that public awareness can be a great tool in the longer term as with litter campaigns, which are a simple means of educating people not prepared to let something happen.

The way we've probably got to go with this is to motivate organisations, whether they be angling groups or conservation groups, to push things in the right direction, i.e. to demand that these processes take place and that there be some coordination for them as well, which is something very much lacking. As Bob O'Boyle said, we should take the lead. There are at least 10 organisations with management responsibilities for fresh water in Tasmania, and no one is prepared to take the lead in many cases. Someone does have to take the lead, to say this is how we do need to go, and to develop some sort of strategy. We are short of time and short of resources, but we can get some of the public to demand of the Ministers that something be done, and get someone to coordinate these issues—that is one way we can move forward.

Barbara Richardson suggested that one of the very important matters Wayne Fulton had raised was about dealing with government agencies and shifting their attitude. What we should be aiming at is to make them think differently. External pressure in the community can be harnessed to help change attitudes, and strategic planning processes in Fisheries agencies can also reflect the importance of this component in fisheries management.

Peter Jackson again emphasised the importance of data. In Queensland, where Water Resources is listening to Fisheries, they will build fishways. In fact there is legislation to force people to build fishways; but there is not yet a complete data set to build proper fishways, and this can lead to enormous problems. At the moment a million dollars is being spent on two fishways, and as Martin Mallen-Cooper will tell, it is not certain whether one of those fishways will work or not. So we have the goodwill there, we've got the cooperation, and we have *some* data, but if we do the wrong thing we can set everything back twenty years. We need to have the data as well as knowing what the threats are.

Wayne Fulton agreed with Peter Jackson. In Tasmania there is no legislation for fishways, or even the right to talk about water allocations. There is no legislation to cover that at the moment. So they have just had to muscle their way in to be able to debate water allocations with some of those other groups without any legislative backing at all.

Mick Olsen wanted to know whether the emphasis in Tasmania is now on the native fish rather than trout.

Wayne Fulton pointed out that this is not the case. There has been a lot of work done in recent years on the native fish, but there are still very strong recreational fisheries based on the salmonids, and a need to balance those interests. Certain elements of the public, if they had their way, would go even more towards the salmonid side. However, it is the view of the Inland Fisheries Commission that we need the balance without that being necessarily the view of the wider public.

Chairperson Barbara Richardson suggested that, as this was the last session before the General Discussion, the focus should now be on what information managers needed in order to manage and make decisions to improve Fish Habitat Management.

Rob Lewis was interested in Bob O'Boyle's reference to transferable pollution property rights. When transferring rights, does that take into account the different or similar capacities of different water masses, or can they just be transferred across just any water masses; or do you approach it on the irrigated watershed management strategy?

Bob O'Boyle replied that he had added that as a thought for discussion, but in fact they were not using those rights now at all. He knew that it had been used in other fields, certainly in air pollution, and maybe it is something that could be considered.

Barbara Richardson commented to Rob Lewis that NSW is having to face that now, because the NSW EPA has new legislation which provides for pollution rights; and the Fisheries agency has to develop a response to that as managers. But it is one of those issues on which there isn't much information at all. However, some information is being collected on some of the problems with bioaccumulative materials that have been licensed for discharge and give cause for concern. Other components, such as nutrients, are seen as potentially suitable for management by property rights at this point in time. Priority is not being directed towards those problem chemicals that are persistent at this stage.

Karen Edyvane responded to Barbara Richardson's plea for tools of management by refocussing the debate on to ecological modelling, which, in common with Bob O'Boyle, is assuming priority in her thinking. That is, that when you are looking for tools of management, particularly at a regional level in an ecosystem, it is of the utmost importance to model ecosystem responses. Australia certainly has a number of initiatives underway at the moment, and she invited Norm Hall to comment on some of the initiatives in ecosystem modelling and the role it can play, not only as a management tool but also in directing research through concentrating on processes.

Norm Hall responded first of all with some background.

The problem of effluent disposal for the northern Perth metropolitan region is growing. We have an increasing population. Current disposal techniques involve secondary treatment, followed by piping the waste water into Marmion lagoon, a fairly narrow coastal lagoon bounded by an offshore reef, where the average depth of water is about 10 metres. While it is considered that the flushing time of the lagoon is of the order of about 1 to 2 days in the region of interest, there is concern that the nutrients released into the lagoon from the outfall may simply slosh up and down along the coast, rather than exchanging and mixing with oceanic water from outside the bounding reef.

Some years ago, the Western Australian Water Authority constructed the first outfall within the Marmion lagoon. Subsequently, a marine park was established within the same region. The Environmental Protection Authority, EPA, approved the initial outfall, and agreed to a subsequent increase in the volume of waste water to be discharged, conditional on the Water Authority establishing that the impact of the increased nutrient load would be acceptable. Of concern are not only health, and aesthetic qualities, but also the potential changes that might occur within the plant and animal communities.

The EPA required the Water Authority to assess the assimilative capacity of the northern metropolitan waters. The Water Authority faced the need to replicate the pipeline into the lagoon (or the northern suburbs would be awash with sewage), and to increase the flow of effluent. The EPA had constrained the total amount of nutrients that might be released, and projected increases in effluent from the region suggested that by 1995 the nutrient load would reach the EPA specified upper limit. It was therefore urgent that a study be undertaken to assess the assimilative capacity of the lagoon, in order to determine whether further increases in the nutrient load might be permitted, or whether alternative disposal methods might need to be considered.

Nutrients flowing into Cockburn Sound, south of Perth, had resulted in the excessive growth of epiphytes on seagrass leaves, causing the death of the seagrass through the effects of shading. A similar impact had been seen at Princess Royal Harbour near Albany in the south-west of the State. In both cases seagrasses had been lost. The EPA's principal concern for the Marmion lagoon was that the enhanced levels of nutrients might result in the loss of seagrass, with an associated impact on the fish communities of the area. Further outfalls are also planned to cater for the growth of the city. The cumulative impact of the nutrient added to coastal waters by these additional planned outfalls must also be assessed.

Initial proposals to study the physical, chemical, and biological processes operating within the Marmion lagoon, in the vicinity of the Beenyup outfall, lacked integration and appeared excessively expensive. The Water Authority decided that the first thing to do was to bring in a modelling team, and to run a workshop in order to determine the features and processes that required study. It was hoped that this would result in a more cost-effective study.

Professor Carl Walters, from the University of British Columbia, was invited to run such a workshop. This was held at Perth in December 1991. During the workshop, a barotropic model was set up to describe the physical and chemical aspects of the system. Previous studies had suggested that water movement within the system was wind driven. Baroclinic, or temperature and density related, effects were considered of lesser importance. Following on from the workshop, John Hunter, from CSIRO at Hobart, has extended and improved the model of the physical process.

After calculating the movement of water within the system, the chemical processes were calculated. Then, with an understanding of the concentration of chemicals throughout the system, the processes of primary production were examined. Finally, components were added to the model to describe some aspects of secondary production, by including filter feeders, detritivores, and grazers. Little information was directly available for many of the parameters required within the model for the seagrasses, macro algae, and epiphytes within the Marmion lagoon, so values were selected from the literature. A detailed understanding of the processes of growth and mortality of the grazers, filter feeders, and detritivores was completely lacking, and subjective estimates were supplied for the parameters required to describe these processes.

The workshop was successful in bringing together available data, and producing discussion and interaction between the various groups involved in the study. While each group is still in competition for the available funds, there is now considerable interaction between the various studies that are proceeding. The original model is being modified and extended, but provides a framework and basis for all the separate studies. It focuses the work being undertaken, and forces the integration and critical assessment of the information collected. The resulting model is intended to be general in nature, with the facility for it to be applied to other coastal areas by changing the description of bathymetry, outfalls, and habitat.

When the original study began, it was thought that little information was available. In fact, when the modelling process began, and data from earlier studies were collected, a considerable volume of information surfaced. These data are now being brought together within a Geographical Information System.

In summary, the approach has been effective in bringing available information together, in a form readily accessible to managers and scientists. A modelling framework was used to integrate the available data, and assisted in the identification of areas where inadequate knowledge existed. Predictions from the model are being made and tested against observations from the real system. The process of modelling has resulted in the interaction of the researchers involved and the establishment of a common goal, and appears to be working well in both focussing the researchers' minds on the impact of nutrient enrichment within Marmion lagoon and in using their understanding to predict the possible changes that might occur. It should be regarded, however, as only the first step in an ongoing research effort to understand the processes operating within this system. The system is now the subject of a very intensive study which is expected to terminate in 1995.

There is also a very successful plan in place for control of oil spills along the W.A. coast. By identifying the resources at risk, and with a broad understanding of the physical oceanography and the biological systems concerned, a plan to handle oil spills at various locations was formulated. The plan has been tested on several occasions and appears to work well, although its effectiveness in protecting the environment has yet to be tested by a major oil spill.

Barbara Richardson then displayed (on overhead) the following suggested list entitled "Data Needs".

- Habitat information base—what's there and what's happening to it?
- Data on habitat utilisation:
  - critical habitats a priority
  - juxtaposition of habitats and interaction
  - critical links between habitats
  - species interaction and by-catch effects
- How do we 'value' habitats?
- Impacts of harvesting activities:
  - species catch data to include environmental data
  - data on impacts of gear type and operation on habitats
- Effectiveness of protected areas
- Collaborative research—multi-disciplinary/ strategic vs tactical
- Monitoring design and data source

- Assessment of other human activities
- Long-term data sets

In the discussion which followed, Jenny Burchmore stressed the importance of habitat mapping and inventory to management, and the need for it to be repeated and continued.

Sandy Morrison further underlined the importance of monitoring the effects of anything which is done e.g. fish ladder construction. Such feedback is a vital part of the learning experience from case studies.

Jenny Burchmore agreed, but with the reservation that monitoring is not enough. In the event that an undesirable impact is revealed by monitoring, the required management strategies need to be developed.

George Paras wanted to place more emphasis on amelioration. Alternative strategies of demand management are needed e.g. for reducing water use. We should be looking towards the long term and applying what is already known about impacts, rather than accepting the *status quo* as being good enough.

Hugh Cross urged the need for more of a vision, through which to target the decision makers, and encapsulated the three major obstructions demonstrated from Jenny Burchmore's case examples as legislation, final decision makers and the lack of information. Barbara Richardson challenged the group on how to set that vision.

Murray MacDonald believed that the vision would be set by strategic planning and informed community debate - a strategic system framework, involving the assessment of impacts of all other types of human activity.

In closing the session, Barbara Richardson urged participants to consider the issues of management through amelioration, enhancement and conservation and have their thoughts ready for the final discussion.

GENERAL DISCUSSION

Session Chairperson:　　P.C. Young
Session Panellists:　　　R.K. Lewis
　　　　　　　　　　　　　R.K. Gehrke
　　　　　　　　　　　　　B.E. Pierce
　　　　　　　　　　　　　C.M. Macdonald
　　　　　　　　　　　　　R. Reichelt
　　　　　　　　　　　　　B.A. Richardson
Rapporteur:　　　　　　P.R. Last

GENERAL DISCUSSION

Chairperson: P.C. Young
Recorded by P.R. Last

CSIRO Division of Fisheries
GPO Box 1538
Hobart TAS 7001

Chairperson's summary

By way of a brief review of the proceedings of this Workshop over the past few days, I will take five minutes to identify a few of the critical issues that emerged as I saw them and then ask each of the Session Chairs to briefly review the issues for a few minutes. I will then seek to draw the Workshop together to attempt to synthesise a cogent view of the topic "Sustainable Fisheries through Sustaining Fish Habitat".

We started yesterday with an illuminating keynote address by Stan Moberly who clearly identified a number of critical issues in America. These broke down into the need:

1) for scientists to inform users and decision-makers of the value of habitat;

2) to develop better administrative arrangements for managing habitats that are impacted by cross-state or even cross-country activities; and

3) for scientists to have a key role in bringing the conflicting resource uses together for strategic planning.

Following Stan Moberly's address, we examined a manager's view of fish habitats. During this session the central theme that emerged was that there were competing uses of the ecosystem many of which will impact on habitats and fisheries.

The 4 pertinent issues became:

1) Who is going to control the use between conflicting users?

2) How do we minimise the effects of conflicting use on habitat?

3) The need to identify how these effects on habitat affect fisheries.

4) To do 1 – 3 we need to develop appropriate research and monitoring strategies to incorporate multiple use so that:

   (a) decision-makers can develop planning strategies; and

   (b) the data are there for scientific advice.

Turning to the relationship between organisms and environment, we were stimulated by the concept that although we really know very little about the functioning of habitats, we are making sweeping announcements of their value and, following that, have incorporated strategies to modify habitats by engineering and bioremediation. It became clear during the discussion that although we are modifying habitat to produce the "ideal climax", there appeared little agreement as to what was the "ideal climax" nor if we were sure about how to achieve it.

Turning to the organism/environmental relationships, we had some problems defining key factors to fishes. We were fairly sure about freshwater where the notion of limited and critical factors was suggested to develop the concepts of ecophysiological controls and critical use of habitat. However, as we moved farther towards the marine area the precise role of critical habitat became clouded. We identified that habitat is important but don't yet have a really good definition of the importance. We then need to better define our goals. As scientists we want to know more, but as managers we need to be ready to summarise our present knowledge for guidelines.

We must determine what is the important scale for defining and protecting habitat, and although we don't know enough, we must produce a reductionist view. This will be enhanced by the role of the modeller who will help us to define what are the critical elements still unknown about the processes to sustain this minimum necessary knowledge.

Overall, by the end of the day we come back to the issue that we already know enough about some habitats to be at the stage where we must persuade people to understand what we know and to share the conclusions. This might be helped by linkages between professional societies such as the Institution of Engineers or by other initiatives such as a "Status of the Aquatic Habitats" report.

Day 2 started off with the impacts of human activities on habitat and fisheries and discovered that most of the things we summise about the relationship of habitat and fisheries are based on inferences rather than experiments. We were treated to the results of a long term experience which did investigate these relationships. However, because of the cost in dollars and time, it was suggested that perhaps we should now take our best guesses and actively seek to insert them into the decision-making process.

We established that it was important to identify the activities which cause change that impacts on habitats critical to fish communities.

When looking at the alternative uses of habitat, we heard that we may need to set up broad-based refugia from consumptive use, but that an alternative methodology would be to seek to develop the concept of value-adding to habitat use in its broadest sense. This again emphasised the importance of communication, institutional reform, and managing strategically for all users.

We then turned to this concept of management in terms of ameliorating effects, enhancing and conserving habitats.

We heard of the approach to conserving habitats by defining the habitats as functional units, using explanatory models to understand the ways that proposed activities may impact on them, and the need for appropriate regulation and data collecting limited to decision support systems. There may be some benefit in considering the purchase of "pollution rights".

Amelioration and enhancement of habitats is more difficult. We lack the knowledge of how to restore habitats, and at this stage are preparing experiments to work out how to do it. In view of this, what is the *real* compensation for habitat destruction? How are we going to develop the information required to improve habitats and ameliorate effects?

Session Chairpersons' summaries

Andrew Staniford, deputising for Russell Reichelt, commenced with a summary of Session 6 which covered the alternative uses of aquatic habitats. Two proposals were put forward: a regulatory approach based on marine parks or marine development zones as a method of managing the marine ecosystem; and a value-added approach which tries to identify the extra value that can be achieved by changing the allocation of the resource between the competing users, as well as by changing their levels of activity. He stressed that the two approaches are closely linked. Together they form important tools in establishing priorities for strategic planning and determining which areas are important for certain activities. The main advantages of an economic approach are in providing some sort of measurable, objective criteria for comparing uses, and assessing how changes in the way a resource is used will affect the community.

Murray MacDonald, in summarizing Session 5 on the impact of human activities on habitat and fisheries, felt that the discussion had focussed on two major issues. Firstly, with few exceptions (ie. in freshwater and the most accessible environments where the best information is available), evidence suggests that linkages are circumstantial rather than proven by cause and effect type evidence. Secondly, the role of scientists in getting the message across regarding the importance of habitats needs to change. The traditional role of scientists, in conducting research and monitoring type studies to provide detailed information and expert advice on habitat conservation and management, is still a legitimate and very important role but we have to broaden our view of our roles and start thinking of ourselves also as purveyors of information in a much more simplified form.

Bryan Pierce then provided his perspective on Session 4 which dealt with the key variables and broad-based issues that affect organisms and environmental relationships. He concluded that habitats were only unimportant in the deepwater/pelagic zone, and that inshore the priorities diminish progressively from inland through to coastal habitats. A model of the adaptive management process using the key factors was seen to be a "n-dimensional headache" because we cannot even determine general key factors. An educational role was also identified as being critically important.

Peter Gehrke summarized the results of Session 3 covering case studies of organisms and environmental relationships, during which a number of the processes for sustaining fish habitat were addressed. While he had originally believed that our knowledge of fish habitat was in some way proportional to the number of people researching each of the various habitats (ie. that estuarine and coastal systems had been worked to death while freshwater habitats were less well understood), he felt that this viewpoint had been reversed during this workshop. Debate focussed on the need to supply information quickly to address the present need for habitat amelioration, as against the more traditional research response which was seen as causing only more delays before providing "answers". Gaps between managers and researchers need to be filled to provide some urgency of response to habitat issues.

A triage response to avoid time-wasting was proposed with a line of ascending priority from habitats that don't need any real sustenance other than protection, through those that need some form of active sustenance, to those that require totally rebuilding. Model driven research, which feeds the best available information to managers but also highlights the knowledge gaps for researchers, was preferred to piecemeal and fragmentary data fed intermittently to managers.

Peter Gehrke believed that this last approach had led to potential problems in Australia because of a tendency to concentrate on system ecology or ecological questions without, in many cases, evaluating details of the processes driving the systems. "By ignoring some of the finer points of how the processes work we run the risk of trying to run before we can crawl when it comes to managing habitats - we end up getting lost in trying to repair and maintain their virginity". A coordinated modelling effort must be established to provide some form of expert system that can advise managers and at the same time refine the hypotheses needed to direct research. He concluded by raising the issue of the effects of global climate change on fish habitat.

A manager's view of fish habitat (Session 2) was then summarized by Rob Lewis. He believed that the consensus view is that we want to see the habitat managed successfully and we all want to contribute to that outcome. Session 2 had identified a need to address the issues, such as competing use, impacts and demands, which cannot be handled solely by fisheries interests.

A key role for ASFB was also identified despite differing views on specific issues within the membership. This group, with the expertise and the will, was considered to be capable of taking a lead (ie. harnessing, coordinating, and promoting these actions). There is a need to coordinate the scientific component, communication, and the planning and application stages with appropriate strategies at different levels for different audiences. Finally, he observed a widespread emotional commitment to prevent further habitat decline or at least replacement, and to look beyond maintenance alone and consider habitat amelioration.

Before Barbara Richardson's summary of Session 7 on the amelioration, enhancement and conservation issues of management, Peter Young suggested the need for focussing the subsequent discussion and invited Barbara Richardson to present additional overheads considering the focal issues.

Barbara Richardson commenced by pointing out that much of our effort is directed towards firefighting, largely unproductive, and needs to be focussed on a strategic planning process. She acknowledged that the Society had the resources and the commitment and proposed a more structured visionary approach to be outlined in the workshop proceedings. Within the planning process, she felt that we need a clear definition of goals, to consider all the possible strategies available, to develop performance indicators, and most importantly to "look at where we are going and see if we are actually getting there".

She believed that the strategies identified during the workshop had merit, highlighting the entrepreneurial experiences of some participants. As an example of integration, she referred to Bob O'Boyle's experiences in Canada. She proposed an adaptive management strategy and drew attention to several other issues discussed: identification and consideration of problems; the inadequacy of some existing legislation; the possible inadequacy of our information base and future data needs; and the importance of modelling as a strategy for focusing on our information needs, setting of research priorities, and in the design of monitoring programs. She referred to a problem identified by Jenny Burchmore relating to our influence in the decision-making process, and felt that many stakeholders affect these decisions and the Society needs to play a more active role.

A list of the major issues identified by an informal working group the evening before was shown (Tables 1 and 2). Barbara Richardson pointed out that the list was largely unstructured but contained many issues for general consideration. She reiterated the sentiments of other participants that although many of the habitat issues have been discussed for 20 years we haven't achieved much to date. She felt that a current situation statement outlining the issues, and our views concerning possible action plans, should be included in the proceedings. This statement should also highlight the importance of fish habitat management. She outlined some focal questions that might be considered during the drafting process: whether there should be more integration between traditional fisheries management and other ecosystem users; the need for more research and the identification of issues considering data requirements; extracting from managers around the country their perspective of key issues and pressures; and the need to focus scientists' priorities for research and management.

She suggested producing a synthesis of these viewpoints but acknowledged that doing so might not be realistic within the time remaining. The issue was left open for the chairperson to decide but she felt that an outline of current key issues could be used as a yardstick in the future. This would also highlight the next step, particularly if drafted as a situation and action statement, which she suggested might need to be constructed by a drafting committee.

Other important issues included the desirability of having a "state of fish habitats" assessment document with annual upgrades. This document, possibly enlisting broader community help, would complement impending State legislation on environmental reporting. She recommended the formation of an ASFB sub-committee to meet annually with the responsibility of implementing workshop outcomes and reviewing the progress of fish habitat management. She proposed the establishment of an annual newsletter covering fish habitat issues and forming closer links between ASFB and government agencies. She felt the Society could take a leading communication role by convening a national workshop including representatives from all groups whose activities impact on fisheries resources together with conservation groups and government agencies. Society members could also create an awareness of fish habitat by publishing in popular scientific journals such as "Geo" and "Search". A news release could be produced from this workshop drawing attention to the major issues identified.

Peter Young, however, cautioned that it is easy to set work up but it is sometimes harder to find someone to do it. He also pointed out that fisheries managers are now recognising the importance of habitat and suggested that ANZFAC (formerly Standing Committee on Fisheries) may form a committee to monitor habitat in the near future. He then opened the session for question time and invited the president, Julian Pepperell, to suggest where we could progress after the discussion.

Discussion

Don Hancock was concerned about the general viewpoint that "not much had been done". He felt that a great deal had been done but probably not quickly enough. A great deal of work was being done independently by individual States but there seems to be no national forum for the discussion of technical issues, and most environmental committees are forced to work to administrative guidelines. As an example, he referred to his experiences as a member of the Fisheries Pollution Committee (FPC) of the then Standing Committee on Fisheries. He suggested that the Society should investigate the activities of and provide input into such committees. He challenged the perception that there was little scientific communication with the fishing industry or the general public. He believed that communication ranged from quite good to exceptional and that we should be encouraged with our efforts rather than discouraged. He advocated greater communication between the States and referred to the example cited earlier by Jenny Burchmore of the oil pollution atlas.

Peter Young commented that the FPC and affiliates do consider national issues and the CSIRO's Division of Fisheries provides the secretariat for that committee.

Ross Winstanley supported this view, stating that whereas the FPC addressed a narrow field of issues 15 years ago, the committee now displayed a changing emphasis, focusing more on the habitat and environmental aspects of fisheries management. He commented on Peter Young's prediction that ANZFAC may establish a sub-committee to investigate fish habitat issues in relation to fisheries by stating that it was long overdue.

Duncan Leadbitter reiterated the point about the type of information needed and the need to approach the relevant groups for input. In his capacity as a member of NFIC's environment committee, he claimed that an informal approach from ASFB to collaborate on habitat issues would be welcomed. He also offered to coordinate a small working group to draft guidelines for environmental impact assessments.

Peter Young then raised the issue of an inventory of critical habitat. Very good examples of this kind of data have been collected in the USA and some other countries. He sought the views of the meeting and raised a number of related questions: should an inventory be constructed; if so, on a national or State by State basis; what information is available and is it archival; should we approach Environmental

Resources Information Network (ERIN) or National Resources Information Centre (NRIC) to get it onto a Geographical Information System (GIS)?

Murray MacDonald felt that the Society ought to be taking a higher profile in bringing these issues before the general public, as well as the legislators and managers. He suggested that part of the costs of convening a conference and making statements about the state of the marine environment could be met externally by dovetailing the Society's interest with current national initiatives. Two such initiatives are the development of a national "state of the marine environment" report process and the development of a national conservation strategy. The "Ocean Rescue 2000" scheme was suggested as a possible funding source and co-convenor of a conference to discuss these issues.

Roland Pitcher proposed an alternative cost-effective communication method involving the ABC's "Survival" program. He outlined possible contents of episodes and suggested that the ASFB president contact the producers of "Survival" to canvas the issue.

Stewart Frusher, giving the example of his own work on rock lobster, expressed concern at not be able to view changes to habitat beyond diving depths (ie deeper than 30 m). He alluded to the high cost factor of studying habitat at greater depths. His sentiments were shared by Peter Young.

David Smith strongly supported the need to prepare a current situation report together with a separate document on the state of fish habitat. He felt that these issues should be handled by a resurrected habitat sub-committee because the Workshop forum was too large and unwieldy. Peter Young supported these views referring to the highly respected Endangered Fish sub-committee which is the major source of advice in this area to the regulators of relevant fisheries. He stressed that the influence of ASFB shouldn't be underestimated.

Norm Hall, readdressing the issue of an inventory, stated that to convince decision-makers you need performance measures to gauge the effectiveness of habitat protection strategies. He commented that most bureaucracies set up committees when a problem arises whereas researchers traditionally collect more data to study the problem. Now, both groups see the need to develop models of these systems to help identify suitable management options. As a consequence, the various bodies responsible for managing aquatic habitats will require more integrated types of studies with a focus on modelling the underlying process. Such a focus will certainly help bring the data together. However, he cautioned that the modelling exercise is not a panacea and does have serious limitations in the present climate where data sets are incomplete. He also alluded to problems of communication and bringing regulatory bodies together.

Peter Young took up this point and raised the issue of integrating catch and management, the effectiveness of which is entirely dependent on the good will of the various regulatory bodies involved. In these instances, the modelling approach acts "as a facilitator to get pig-headed people to become less pig-headed".

Dianne Hughes commented on the size of goals suggested, intimating that they were probably overambitious. She felt, referring to issues raised by Barbara Richardson, that it might be more sensible to set smaller, more achievable goals.

John Koehn agreed with most of the ideas that Barbara Richardson had put forward and believed that the workshop should take Stan Moberly's advice and be pro-active. ASFB, being immune from some of the problems confronting us within our various agencies, could become Australia's leading voice. He stated that workshops should have a bottom line, setting out priority needs and areas. While uncertain about its worth for marine habitats, he believed it possible for freshwater and wished to see action taken at this workshop.

Martine Kenlock followed up the suggestion by Roland Pitcher by offering to contact the producer of "Survival" on behalf of the Society.

Peter Young conceded that it was impossible to progress further given the size of the forum and invited Julian Pepperell, as president of ASFB, to give his perspective on where the Society should go from here.

President's summary

Julian Pepperell agreed with the panellists' view of the Society's role and endorsed most of their ideas, stressing the need to be independent. He addressed the issue of conflicting interests, which sometimes occurs between the Society and the Institutional roles of members, but quoted Stan Moberly's experiences with the American Fisheries Society which has been very successful in providing an independent voice in the USA. He strongly endorsed re-establishment of the habitat sub-committee and invited Barbara Richardson, as the newly appointed chairperson, to appoint additional members. He agreed with most, if not all, of the action plan provided by Barbara Richardson in Tables 1 and 2 and suggested that a copy should be made available to members at this workshop. He felt that the Society had the expertise to provide information on a national scale with members needing to think more broadly beyond their own regional boundaries. The approaches of the National Marine Fisheries Service and the American Fisheries Society, in producing a document on the state of aquatic habitats, were used as role models. He suggested coordinating with other similar interest groups, such as "Ocean Watch 2000", but without compromising the Society's position. The Society's major goal should be to assess the plight of habitat over the *past* 200 years of European settlement in order to warn decision-makers of possible dangers in the *next* 200 years. He stated that ASFB should be in an ideal position to provide advice on key issues to any related sub-committee established by ANZFAC.

The role of this new sub-committee would be similar to that of the Endangered Fish Sub-committee which has an extremely important advisory role.

Peter Young then closed the session and asked Stan Moberly to deliver his summary.

Table 1. Habitat management

- Present situation—Analysis
- What do we need to achieve?

  By – Amelioration → site/system specific

  – Enhancement → site specific

  – Conservation → package of measures to apply to all waters

- How will we know if it's working?
- Data needs
- How can we improve the protection and management of fish habitat?

Table 2. Improving fish habitat protection and management

- Agency profile and effectiveness
- Researchers and managers promote awareness and information
- Collaborative programs with other government agencies or disciplines
- Community programs—involvement
- Media
- ASFB

SUMMING UP

SUMMING UP

S.J. Moberly

Northwest Marine Technology Inc.
PO Box 99488
Seattle WA 98199-0488

My observations over the past two days have lead me to list several remarks and statements I heard. We have "wrung our hands" and "ground our teeth" over the impacts of habitat loss and there seems to be much we do not know or understand about aquatic ecosystems and how they interact. But, during our discussions these past two days, there were several points worthy of note.

In no special order, some of the things I noted are:

- Aquatic habitat is valuable; it is not plentiful, it is fragile and it is disappearing rapidly.
- There are too many people.
- Most people are ignorant about the value of aquatic habitat.
- Change happens and the rate of change often outstrips our ability to learn, understand and react.
- Development will continue to destroy aquatic habitat.
- Some habitats are beyond help and we should "tend to the crown jewels and the walking wounded".
- Fish and the other "critters" need a place to live and our understanding about how ecosystem(s) function must be more complete.
- Covering aquatic habitat with dredge material spoils the habitat.
- Sewage isn't good for sea life nor people.
- We know that the catchment, the rivers, the estuaries and the nearshore marine waters are important to fish.
- Man's activities don't "create" fisheries habitat-mostly man's activities destroy or compromise habitat.
- Man's activities don't cause fish populations to increase-it is mostly the opposite.
- We know that the general trend of most fish populations is downward; we just don't fully understand politically how to reverse the trend.
- We know that if you drag bottom trawls over the sea bed you will kill the plants and animals that live there.
- It is a mistaken concept that politicians lead.
- It is a mistaken assumption that "top decision makers" make good decisions and that it comes naturally.
- Top decision makers make good decisions with good information.
- Top decision makers need timely information.
- Politicians move with consensus.
- We should share our knowledge. our work is fun, its exciting, and we should educate our citizens.

- We are not good at sharing our knowledge outside our own circle.
- We could be better at sharing our knowledge; we must learn how.
- Others care about aquatic resources and habitat too.
- We can identify our allies.
- We can form partnerships—to teach and to learn; relationships are best when both parties are enhanced.
- Not everyone will share our values or our points of view.
- Sometimes the best solution isn't possible.
- Sometimes the best solution is the one that makes everyone equally unhappy.
- It is easier to "walk backwards into the future;" it is more difficult to affect the future and implement our visions.
- Ecologically sustainable might have been present before man intervened.
- Helping to manage human impacts on habitat is our most difficult task.
- The smarter we get the easier the job becomes.
- We need to learn to forgive ourselves for our uncertainties and our shortcomings.
- Multiple use is not synonymous with compatible uses.
- Value comes in dollars but it also is measured in the needs, wants, desires and aspirations of people.

I am sure I missed some points of what we talked about over these past two days but I believe I captured the essence of what transpired. The inescapable conclusion is that we know a lot about the problems with habitat and what to do about it! And we recognize there is much we do not know or understand. But our lack of understanding is mostly our secret! Australians believe fisheries scientists and managers know what is wrong with the fisheries and how to correct it.

There seemed to be consensus that action is better than inaction and that scientists and managers are determined to play a role in helping to shape Australia's fishing future. I stated in my keynote address, that for a nation to maintain successful fisheries they need: national strategies, implementation schemes, adequate money, enforcement, adequate legislation, adequate assessment, and periodic audit to measure progress. You know Australia's fisheries are a valuable, renewable national asset. There is consensus that success is better guaranteed with your attention, your views, your energy, and your knowledge. There is no group more knowledgeable about the fishery resources of the nation than yours.

I think important conclusions have been reached during this workshop. I feel a determination by the participants. And I sense that the ASFB will elect to play an even more active role to have a positive influence on what the future holds for this continent's fisheries. If not you; who? There is no group more capable of leading. I sense you are determined to be Australia's principal champion for its fishing future and I applaud you.

I appreciate very much being invited to participate in this workshop. You live in a fun and wonderful land and these are important times for fish and fish habitat.

APPENDIXES

Appendix 1. Australia's Threatened Fishes 1992 Listing—
Australian Society for Fish Biology.
By P.D. Jackson, J.D. Koehn and R. Wager

Appendix 2. Workshop Program

Appendix 3. List of Participants

APPENDIX 1: AUSTRALIA'S THREATENED FISHES 1992 LISTING— AUSTRALIAN SOCIETY FOR FISH BIOLOGY

P.D. Jackson[1], J.D. Koehn[2] and R. Wager[1]

[1]Fisheries Division, Queensland Department of Primary Industries, GPO Box 46, Brisbane QLD 4001

[2]Freshwater Ecology Section, Department of Conservation and Natural Resources, PO Box 137, Heidelberg VIC 3084

Introduction

Australian fishes were first classified according to their conservation status by the Australian Society for Fish Biology (ASFB) at a Conference on Australian Threatened Fishes in Melbourne in August 1985 (Harris 1987). As a result of this Conference, the Society established a Threatened Fishes Committee to undertake an annual review of the conservation status listing. The Committee has reviewed the listing each year since 1987 and updated listings have been published in the Society's newsletter. The publications of the 1989 listing in Pollard *et al.* (1990) and Ingram *et al.* (1990) comprise the first international publications of the ASFB's conservation status listing and they include some information on reasons for decline of some species together with possible management options. However, an updated listing has not been formally published and no additional information has been included on the threats to Australian fishes and their habitats since 1990. The lack of comprehensive and updated information on threats to freshwater fishes is particularly significant because one of the recommendations of the Conference in 1985 was that the Committee should 'advise on fish conservation matters including management of habitats of threatened species' (Harris 1987).

The Society's decision to hold a "habitat workshop" has provided the opportunity to formally publish an updated listing and to provide further specific details on threats. The Threatened Fishes List was updated by the Committee at the time of this 1992 workshop.

Conservation status classifications

The following conservation status classification was adopted by the ASFB at its conference in 1985 (Harris 1987) and updated at the Society's conference in 1989:

- Extinct—Taxa which are no longer found in the wild or in a domesticated state.

- Endangered—Taxa which have suffered a population decline over all or most of their range, whether the causes of this decline are known or not, and which are in danger of extinction in the near future. (Special management measures required if the taxa are to continue to survive).

- Vulnerable—Taxa not presently endangered but which are at risk by having small populations and/or populations which are declining at a rate that would render them endangered in the near future. (Special management measures required to prevent the taxa becoming endangered or extinct).

- Potentially threatened—Taxa which could become vulnerable or endangered in the near future because they have a relatively large population in a restricted area; or they have small populations in a few areas; or they have been heavily depleted and are continuing to decline; or they are dependant on specific habitat for survival. (Require monitoring).

- Indeterminate—Taxa which are likely to fall into the Endangered, Vulnerable or Potentially Threatened category but for which insufficient data are available to make an assessment. (Require investigation).

- Restricted—Taxa which are not presently in danger but which occur in restricted areas, or which have suffered a long term reduction in distribution and/or abundance and are now uncommon.

- Uncertain status—Taxa whose taxonomy, distribution and/or abundance are uncertain but which are suspected of being Restricted.

Table 1 shows the 1992 Conservation Status Listing.

Summary of change in listing since 1985

Table 2 shows the numbers of fish classified in each category, each year, since 1985. The total number listed has risen from 59 to 72 and perhaps more significantly, the number of fishes listed as endangered has risen from 4 in 1985 to 11 in 1992.

Prior to 1990 no marine species appeared on the listing. That year the grey nurse shark (*Carcharias taurus*) (Vulnerable), the black cod (*Epinephelus daemelii*) (Potentially Threatened) and the great white shark (*Carcharadon carcharias*) (Uncertain Status) were listed and it was noted that the status of the southern bluefin tuna (*Thunnus maccoyii*) required investigation. Since 1990 there have been no further nominations for marine species and threats to these species are not discussed in this paper.

Threats to freshwater fishes

The exact reasons for the decline of many native freshwater fishes are not well known. However, many threatening processes are well recognised and have been documented and discussed by many authors (e.g. Koehn and O'Connor 1990a; Cadwallader 1978; Wager and Jackson in press). In general, these threats apply to threatened and non-threatened species alike and although the urgency for attention generally rests with the former, such threats continue to cause the decline of many species, moving them closer to inclusion on threatened species lists. Being hidden underwater engenders a lack of public understanding of freshwater habitats and ecosystems, and hampers recognition of many of these threats.

The causes of decline of all species listed in categories considered to contain fish under threat (Endangered, Vulnerable, Potentially Threatened and Indeterminate) were assessed from nominations received or expert opinion sought in relation to each species. Threats are listed in order of numerical occurrence in Table 3 and details of the threatening processes for each species are given in Appendix 1A.

The number of threats listed varied between species, ranging up to seven. For three species no particular threats were identified (see Murray hardyhead *Craterocephalus fluviatilis*, swamp galaxias *Galaxias parvus* and non-parasitic lamprey *Mordacia praecox* in Appendix 1A). It is recognised that some threats for other species may remain unidentified due to a lack of knowledge of their biological requirements. The reasons for the decline of a species may not be due to one single threat but often the result of a combination of several factors, and threats may not be uniform throughout the species range or lifecycle. As well as affecting a fish species, many of the threats also have detrimental effects on other aspects of the freshwater ecosystem.

Detrimental interactions with introduced species were considered to be a threat to 20 of the 26 native species listed, with the mosquitofish *Gambusia holbrooki* being implicated in 9 instances. *Gambusia* is widespread throughout mainland Australia (McKay 1989), exhibiting aggressiveness and wide environmental tolerances which enhance its interactions with other species (Lloyd 1989). This species has been linked to the decline of at least 35 species overseas, and although the mechanisms for interactions with Australian species are not well understood, they are likely to include competition for food and space, and predation, particularly on eggs and fry (Lloyd 1989). Interactions with brown trout *Salmo trutta* (7), redfin *Perca fluviatilis* (3), and rainbow trout *Oncorhynchus mykiss* (2), mainly through the process of predation, are major threats to many species, particularly smaller ones (such as *Galaxias* spp.) which are often in otherwise relatively natural habitats. The effects of trout and their exclusion of smaller native fish species has previously been documented (Tilzey 1976; Fletcher 1979; Jackson and Williams 1980; Jackson 1981). In three instances native fish which have been translocated, or in one case have invaded waters outside their natural range, were considered a threat to indigenous species (see Lake Eacham rainbowfish *Melanotaenia eachamensis*, Pedder galaxias *Galaxias pedderensis* and Mary River cod *Maccullochella* n.sp. in Appendix 1A).

The over-riding threats causing the decline of fish species are the destruction and alteration of habitats. These have been listed under a variety of headings. The key habitat variables for freshwater species have been outlined by Koehn (this meeting) and include the amount of water, physical and chemical characteristics of water, instream objects, and the immediate surroundings which influence the water. It is the alteration to one or more of these variables that can cause a decline in fish populations. As each fish species has different habitat requirements, each habitat variable can pose a different degree of threat to each species when that attribute is altered. Very rarely do such alterations cause an immediately recognisable decline in fish numbers (of course toxic spills may), but more often they have subtle effects which may cause a slow population decline.

Instream habitat alteration and destruction due to removal (desnagging, channelisation, excavation, gravel extraction), swamp drainage, flooding after dam formation, cattle trampling, mining, development and forestry activities contributed to the decline of 16 species. The removal of riparian vegetation, which provides shade, food and organic inputs, habitat, filtration of runoff and bank stability was also seen as a major threat (13). In addition to its importance to fish species, riparian vegetation is an essential component of the aquatic ecosystem and its removal has an overall detrimental effect. Sedimentation (from forestry activities, roads, developments, mining, agriculture and catchment erosion) was also seen as a contributing factor in habitat degradation in many cases (9). The importance of instream habitat and riparian vegetation are explored by Koehn and O'Connor (1990b).

Changes to water quality (8 cases) included changes to temperature due to low level releases from impoundments, inputs of nutrients and one case of toxic inputs due to mining.

Reductions in flooding and the seasonal reversal of streamflow from irrigation impoundments, reduced and regulated flows and reductions in the water table (particularly affecting springs) through water extraction were assessed as a threat to seven species.

Four large angling species (see trout cod *Maccullochella macquariensis*, Mary River cod *Maccullochella* n.sp., eastern freshwater cod *Maccullochella* n.sp. and Macquarie perch *Macquaria australasica* in Appendix 1A) were considered to be threatened by overfishing (recreational, commercial and illegal), while the red-finned blue-eye (*Scaturiginichthys vermeilipinnis*) was considered to be under threat due to collection.

Barriers which prevented the upstream migration of fish and the downstream movements and dispersal of larvae were considered to seriously affect two species (see Mary River cod *Maccullochella* n.sp. and Australian grayling *Prototroctes maraena* in Appendix 1A), although the effects of such barriers may well be underestimated due to the lack of knowledge of the movement requirements of many species. A lack of genetic diversity in small populations was also considered to be a threat in itself, and this is likely to apply to several other species.

Conclusion

According to Wager and Jackson (in press) there are about 195 species and subspecies of freshwater fish formally described from Australian waters. A further 22 taxa are currently recognised but have yet to be formally described.

Since European settlement, no species are known to have become extinct although one species, the Lake Eacham rainbowfish (*Melanotaenia eachamensis*), no longer occurs in the wild. Five per cent of the fauna are now classified as endangered (i.e. they are in danger of extinction in the near future) and 13% are under threat.

The major threats to freshwater fishes appear to be interaction with introduced species and habitat degradation (see also Koehn and O'Connor 1990a; Wager and Jackson in press). Threats from introduced species range from competitive interactions for food and space to direct predation. At present, the two species most often implicated are brown trout (*Salmo trutta*) and gambusia (*Gambusia holbrooki*). Whilst total eradication of these species is impractical, localised programs to remove these species in areas containing threatened fishes is a valid management tool.

As outlined above, threats to the habitats of freshwater fishes are many, and threatening processes are often interlinked. Although there remains a paucity of data on the habitat requirements of many fishes and those habitat variables that are predictive of fish community structure, much general knowledge is available (Koehn this meeting). Whilst appropriate research is essential and must be encouraged, programmes to mitigate degradation of freshwater habitats must be initiated now with the best data presently available.

In the case of some of the smaller species that occur in very fragile and restricted habitats e.g. the honey blue eye (*Pseudomugil mellis*) that occurs in coastal wallum swamps and the red-finned blue eye (*Scaturiginichthys vermeilipinnis*) that inhabits shallow artesian soaks, total habitat destruction remains a real possibility. In these cases the protection of specific areas by the development of management plans with local landholder cooperation or the creation of conservation reserves or national parks is a valid management option. However, as illustrated by the case of the Lake Eacham rainbowfish (*Melanotaenia eachamensis*), protection within a national park does not necessarily mean protection from introduced or translocated fishes.

In the case of larger and more widely distributed species, e.g. Macquarie perch (*Macquaria australasica*) and Australian grayling (*Prototroctes maraena*), a total catchment management approach is required to ensure habitat protection. Cooperation from local communities, particularly private landholders, is essential. Measures to reduce catchment erosion and to protect remnant native riparian vegetation and replanting programmes to enhance it where necessary, will be of significant benefit to a large number of threatened and non-threatened fishes alike.

Water extraction and impoundments can have major impacts on threatened fishes. Consideration must be given to the needs of these fishes in the management of water extraction and the control of impoundment outflows. Alterations to low-level offtakes, which may change both water temperatures and water quality downstream, and the inclusions of fishways are required to negate these widespread problems. Controls over snag removal and other forms of instream habitat destruction (e.g. channelisation, stream clearing etc.), are also needed.

The number of species on the conservation status listing has increased from 59 in 1985 to 72 in 1992. More significantly, the number of threatened fishes (i.e. those in the endangered and vulnerable, potentially threatened or indeterminate categories) has risen from 15 to 28. Whilst this may in part be due to more data now being available on some species, it is significant that no threatened species has been removed from the list in the seven years since 1985 as a result of rehabilitation measures. The species recovery programs initiated by the Endangered Species Unit of the Australian National Parks and Wildlife Service may assist in this area (see Wager and Jackson in press).

Recommendations

At its annual meeting at Victor Harbour, South Australia, in 1992, the Threatened Fishes Committee of ASFB outlined a number of areas of concern in relation to the conservation of fishes. These concerns are recorded here in the form of Recommendations.

Taxonomy
The Threatened Fishes Committee is concerned about the number of undescribed taxa that appear on the list (e.g. *Chlamydogobius* sp., *Glossogobius* sp., *Mogurnda* sp. etc) and other groups which have been identified as requiring taxonomic work (e.g. *Macquaria australasica*). Some taxa have remained undescribed since the list was formulated in 1985. Taxa which are of particular concern and which should be given priority are: *Galaxias fuscus*, *Mogurnda* n.sp., *Neosilurus* n.sp., *Chlamydogobius* n.sp. and *Macquaria australasica*. It was noted that the taxonomy of *Maccullochella* n.spp. (eastern freshwater cod and Mary River cod) has been resolved (Rowland in press).

Recommendation. That the Australian Society for Fish Biology encourages the necessary taxonomic work to enable formal descriptions of presently undescribed taxa to be published as a matter of urgency.

Updating of other conservation status listings
A number of other conservation status listings of fish species exist. Most of these use the ASFB listing as their basis (e.g. the Australian and New Zealand Environmental Conservation Council list of Endangered Vertebrate Fauna, the IUCN Threatened Australian Freshwater Fishes List). However, these listings do not keep pace with the annual changes to the ASFB listing.

Recommendation. That the Annual ASFB Threatened Fishes List be forwarded to all relevant agencies compiling conservation status listings with a view to their listings being updated annually in accordance with the ASFB list.

Attention to fish species in Indeterminate and Uncertain Status categories
Those taxa that fall into Indeterminate or Uncertain Status categories are placed there because there are insufficient data to list them in other categories. There is a danger that they will remain there if the necessary data collection is not encouraged. For example, the blind cave eel (*Ophisternon candidum*) has been in the Indeterminate category since 1985.

Recommendation. That the Australian Society for Fish Biology encourages the necessary research work to be undertaken to collect the required data on taxa in the Indeterminate and Uncertain Status categories.

Need to address threats to fishes

There is a requirement for the Society to be proactive in addressing threats to fishes. There is a danger that no action will be taken until a species becomes listed, whereas threatening processes could be addressed before this happens. For example, the Threatened Fishes Committee received a nomination in 1992 to list silver perch (*Bidyanus bidyanus*) as 'Restricted' due to evidence of decline in populations in New South Wales. The Committee did not accept the nomination because silver perch are still *comparatively* abundant in relation to other species currently listed as restricted. However, there is a need to address the processes threatening silver perch now or inevitably it *will* be listed in the future.

Recommendation. That the Australian Society for Fish Biology becomes proactive in addressing threatening processes. The Society's proposed Habitat Committee may be an avenue for this. There should be close liaison between this Committee and the Threatened Fishes Committee.

Acknowledgements

The authors wish to thank J. Harris for his foresight in convening the initial conference in 1985 and establishing the Threatened Fishes Committee. The following people attended the 1992 Committee meeting and contributed to a lively debate; W. Fulton, D. Pollard, P. Unmack, J. Anderson, K. Kukolic, S. Saddlier, M. Mallen-Cooper, B. Pierce and P. Cadwallader. In addition nominations were provided by T. Raadik, H. Larson, K. Bishop, G. Gooley and J. Douglas. Much of the data presented in Appendix 1A was collected while two of the authors were preparing the Action Plan for Australian Freshwater Fishes for the Australian National Parks and Wildlife Service. The Service's financial support and foresight in commissioning this project is gratefully acknowledged.

References

Cadwallader, P.L. (1978). Some causes of the decline in range and abundance of native fish in the Murray-Darling River system. *Proceedings of the Royal Society of Victoria* **90**, 211–224.

Fletcher, A.R. (1979). Effects of *Salmo trutta* on *Galaxias olidus* and macroinvertebrates in stream communities. MSc thesis, Monash University, Clayton, Victoria.

Harris, J.H. (Ed.)(1987). *Proceedings of the Conference on Australian Threatened Fishes*. Australian Society for Fish Biology and Department of Agriculture NSW, Sydney, 70pp.

Ingram, B.A., C.G. Barlow, J.J. Burchmore, G.J. Gooley, S.J. Roland and A.C. Sanger (1990). Threatened native freshwater fishes in Australia - some case histories. *Journal of Fish Biology* **37**(Supplement A), 175–182.

Jackson, P.D. (1981). Trout introduced into south-eastern Australia: their interaction with native fishes. *Victorian Naturalist* **98**, 18–24.

Jackson, P.D. and W.D. Williams (1980). Effects of brown trout, *Salmo trutta* L. on the distribution of some native fishes in three areas of southern Victoria. *Australian Journal of Marine and Freshwater Research* **31**, 61–67.

Koehn, J.D. and W.G. O'Connor (1990a). Threats to Victorian native freshwater fish. *Victorian Naturalist* **107**, 5–12.

Koehn, J.D. and W.G. O'Connor (1990b). *Biological Information for Management of Native Freshwater Fish in Victoria*. Government Printer, Melbourne. 165pp.

Koehn, J.D. (this meeting). Freshwater fish habitats: Key factors and methods to determine them.

Lloyd, L. (1989). Ecological interaction of *Gambusia holbrooki* with Australian native fishes. In: Pollard, D.A. (ed). Introduced and translocation fishes and their ecological effects. *ASFB Workshop Proceedings* **No.8**, 94–97.

McKay, R.J. (1989). Exotic and translocated freshwater fishes in Australia. In: De Silva, S.S. (ed.) Exotic aquatic organisms in Asia. *Proceedings of the Workshop on Introduction of Exotic Aquatic Organisms in Asia. Asian Fisheries Society Special Publications* **3**, 21–34.

Pollard, D.A., B.A. Ingram, J.H. Harris and L.F. Reynolds. Threatened fishes in Australia—an overview. *Journal of Fish Biology* **37** (Supplement A), 67–78.

Rowland, S.J. (in press). *Maccullochella ikei*, an Endangered species of Freshwater Cod (Pisces: Percichthyidae) from the Clarence River System, New South Wales and *M. peelii mariensis* a New South Wales subspecies from the Mary River System, Queensland. *Records of the Australian Museum.*

Tilzey, R.D.J. (1976). Observations on interactions between indigenous Galaxidae and introduced Salmonidae in the Lake Eacumbeme catchment, New South Wales. *Australian Journal of Marine and Freshwater Research* **27**, 551–564.

Wager, R. and P.D. Jackson (in press). *The Action Plan for Australian Freshwater Fishes.* Australian National Parks and Wildlife Service, Canberra.

Table 1. The 1992 Conservation status listing of Australian fish species

Category	Scientifc name	Common name
Extinct	No species	
Endangered	*Chlamydogobius n.sp.	Elizabeth Springs goby
	Galaxias fontanus	Swan galaxias
	Galaxias fuscus	barred galaxias
	Galaxias johnstoni	Clarence galaxias
	Galaxias pedderensis	Pedder galaxias
	Maccullochella macquariensis	trout cod
	* Maccullochella n.sp.	eastern freshwater cod
	* Maccullochella n.sp.	Mary River cod
	Melanotaenia eachamensis	Lake Eacham rainbowfish
	Nannoperca oxleyana	Oxleyan pigmy perch
	Scaturiginichthys vermeilipinnis	red-finned blue-eye
Vulnerable	Carcharias taurus	grey nurse shark
	Galaxias tanycephalus	saddled galaxias
	* Mogurnda n.sp.	Flinders Ranges gudgeon
	Nannoperca variegata	variegated pigmy perch
	Pseudomugil mellis	honey blue-eye
Potentially Threatened	Craterocephalus dalhousiensis	Dalhousie hardyhead
	Craterocephalus fluviatilis	Murray hardyhead
	Craterocephalus gloveri	Glover's hardyhead
	Epinephelus daemelii	black cod
	Galaxias parvus	swamp galaxias
	Galaxiella pusilla	dwarf galaxias
	Mordacia praecox	non-parasitic lamprey
	* Neosilurus n.sp.	Dalhousie catfish
	Edelia obscura	Yarra pigmy perch
	Prototroctes maraena	Australian grayling
Indeterminate	Macquaria australasica	Macquarie perch
	Ophisternon candidum	blind cave eel
Restricted	Cairnsichthys rhombosomoides	Cairns rainbowfish
	* Chlamydogobius n.sp.	Dalhousie goby
	Craterocephalus amniculus	Darling River hardyhead
	Craterocephalus centralis	Finke River hardyhead
	Craterocephalus helenae	Drysdale hardyhead
	Craterocephalus lentiginosus	Prince Regent hardyhead
	Craterocephalus marianae	Magela hardyhead
	Galaxias rostratus	flat-headed galaxias
	Galaxiella munda	western mud minnow

Table 1. The 1992 Conservation status listing of Australian fish species (continued)

Category	Scientifc name	Common name
	Galaxiella nigrostriata	black-striped minnow
	* Glossogobius n.sp.	Mulgrave goby
	Hannia greenwayi	Greenway's grunter
	Hephaestus epirrhinos	long-nose sooty grunter
	Hypseleotris aurea	golden gudgeon
	Kimberleyeleotris hutchinsi	Mitchell gudgeon
	Kimberleyeleotris notata	Drysdale gudgeon
	Leiopotherapon aheneus	Fortesque grunter
	Leiopotherapon macrolepis	large-scale grunter
	Lepidogalaxias salamandroides	salamanderfish
	Melanotaenia exquisita	exquisite rainbowfish
	Melanotaenia gracilis	slender rainbowfish
	Melanotaenia pygmaea	pygmy rainbowfish
	Milyeringa veritas	blind gudgeon
	Mogurnda adspersa	southern purple-spotted gudgeon
	* Mogurnda n.sp.	false-spotted gudgeon
	Paragalaxias mesotes	Arthur's paragalaxias
	Pingalla midgleyi	Midgley's grunter
	Scleropages leichardti	Saratoga
	Scortum parviceps	small-headed grunter
	Syncomistes rastellus	Drysdale grunter
Uncertain Status	*Ambassis elongatus*	yellowfin perchlet
	Carcharodon carcharias	great white shark
	Cinedotus froggatti	Froggatt's catfish
	Hypseleotris ejuncida	slender gudgeon
	Hypseleotris kimberleyensis	Barnett River gudgeon
	Hypseleotris regalis	Prince Regent gudgeon
	* Neosilurus n.spp.	undescribed tandans
	Pingalla gilberti	Gilbert's grunter
	Scortum hillii	leathery grunter
	* Scortum n.sp.	Angalarri grunter
	Syncomistes kimberleyensis	Kimberley grunter
	* Tandanus n.sp.	Bellinger River tandan
	Thryssa scratchleyi	freshwater anchovy
	Toxotes oligolepis	big-scale archerfish

Species identified by the Committee at its 1990 meeting as requiring investigations of their status were: *Thunnus maccoyii*   southern bluefin tuna

* Denotes taxa where formal taxonomic description has not been published but where listing is essential because of concern over their conservation status. Early formal publication will be encouraged to resolve their taxonomic status.

Table 2. Numbers of fish in each category of Australian threatened fish species

Classification	1992	1991	1990	1989	1988	1985
Extinct	0	0	0	0	0	0
Endangered	11	9	8	6	5	4
Vulnerable	5	6	6	5	5	4
Potentially Threatened	10	7	5	4	3	5
Indeterminate	2	2	2	2	2	2
Restricted	30	32	30	31	33	29
Uncertain Status	14	13	13	16	16	15
Total	72	69	64	64	64	59

Table 3. The number of Australian threatened freshwater fish species affected by each threatening process. A category may be the result of more than one process (Appendix 1A)

Threatening processes	Threatened species category				
	Endangered	Vulnerable	Pot. Threat.	Indetermin.	Totals
Interactions with introduced species	10	3	6	1	20
Instream habitat removal/destruction	8	2	6		16
Riparian veg. removal	8	2	3		13
Sedimentation	5	1	1	2	9
Water extraction flow regulation	3		4	1	8
Reduced water quality	1	2	3	1	7
Overfishing/collection	4	1		1	6
Barriers	2		1		3
Loss of genetic diversity	2				2
Unknown			3		3

Appendix 1A. Summary information for each threatened fish (i.e. Endangered, Vulnerable, Potentially Threatened and Indeterminate categories)

Endangered

Elizabeth Springs goby, *Chlamydogobius* n.sp.

Current distribution: Only in Elizabeth Springs complex, central Queensland.

Habitat: Shallow waters of artesian springs.

Threatening processes: Reduction in water level due to lowering of water table by water extraction through artificial bores; habitat destruction due to trampling by domestic stock.

Swan galaxias, *Galaxias fontanus* Fulton (1978)

Current distribution: Restricted to five small streams in the Macquarie and Swan River drainage in eastern Tasmania.

Habitat: Slow to moderately fast flowing streams, around woody debris, rocky pools or stream margins.

Threatening processes: Interaction with brown trout (*Salmo trutta*), cannot co-exist with this species, probably due to predation. Habitat alteration due to forestry operations.

Barred galaxias, *Galaxias fuscus* Mack (1936)

McDowall and Frankenberg (1981) consider it a junior synonym of *G. olidus*. Taxonomic status needs to be confirmed.

Current distribution: Restricted to six small streams in the upper reaches of the Goulburn River between Marysville and Mount Howitt.

Habitat: Upper reaches of small, clear flowing streams in mountainous country with gravel or boulder substrates. Preference for pools.

Threatening processes: Predation by rainbow trout (*Oncorhynchus mykiss*) and brown trout (*Salmo trutta*). Habitat degradation and sedimentation from forestry and mining operations and road construction. At least one spill of cyanide from mining operations has occurred.

Clarence galaxias, *Galaxias johnstoni* Scott (1936)

Current distribution: Restricted to the headwaters of the Clarence River, headwaters of Dyes Rivulet and Dyes Marsh and a lagoon in the Wentworth Hills, south eastern Tasmania.

Habitat: Rocky margins of streams and lagoons.

Threatening processes: Interactions with brown trout (*Salmo trutta*).

Pedder galaxias, *Galaxias pedderensis* Frankenberg (1968)

Current distribution: Reported from two small streams draining into Lake Pedder, Tasmania; may still occur in three other small streams in the area, but not recently captured.

Habitat: Slow flowing streams with sandy substrate and abundant cover.

Threatening processes: Interactions with climbing galaxias (*Galaxias brevipinnis*) and/or brown trout (*Salmo trutta*).

Trout cod, *Maccullochella macquariensis* Cuvier and Valenciennes (1829)

Current distribution: Restricted to a few isolated populations in Victoria, New South Wales and Australian Capital Territory.

Habitat: Inhabits both fast flowing and still waters; within streams, occurs in fast flowing water over bedrock, boulder and gravel substrates. Larger individuals inhabit deep holes.

Threatening processes: Interactions with introduced species: brown trout (*Salmo trutta*) carp (*Cyprinus carpio*), redfin (*Perca fluviatilis*) and possibly goldfish (*Carassius auratus*). Habitat degradation caused by desnagging, removal of riparian vegetation, flow regulation in diminishing populations, barriers caused by dam and weir construction, and overfishing.

Eastern freshwater cod, *Maccullochella* n.sp.

Current distribution: Restricted to Nymboida and Mann Rivers in the Clarence River Drainage, New South Wales. Fingerlings have been stocked in the Richmond River Drainage.

Habitat: Clear, rocky streams.

Threatening process: Overfishing, habitat degradation through accelerated catchment erosion, stream siltation and loss of riparian vegetation. Loss of genetic diversity in diminishing populations.

Mary River cod, *Maccullochella* n.sp.

Current distribution: Restricted to a few of the larger tributaries of the Mary River, south eastern Queensland.

Habitat: Found in deeper pools of relatively undisturbed tributaries where fallen timber, branches and boulders provide cover.

Threatening processes: Flow regulation and physical barriers caused by construction of dams and weirs. Loss of native riparian vegetation and siltation from accelerated catchment erosion due to agriculture and forestry practices. Stream channel drainage from sand and gravel extraction and possible interactions with translocated species, golden perch (*Macquaria ambigua*) and silver perch (*Bidyanus bidyanus*). Overfishing.

Lake Eacham rainbowfish, *Melanotaenia eachamensis* Allen and Cross (1982)

Current distribution: No longer found in wild but several captive populations exist.

Habitat: Formerly found in clear, shallow waters at margins of Lake Eacham, often amongst aquatic vegetation, fallen logs or branches. However individuals could be found throughout surface waters of the Lake.

Threatening processes: Probably interactions with translocated species, particularly the mouth almighty (*Glossamia aprion*) but also banded grunter (*Amniataba precoides*), archerfish (*Toxotes chatareus*) and bony bream (*Nematolosa erebi*).

Oxleyan pigmy perch, *Nannoperca oxleyana* Whitley (1940)

Current distribution: Coastal wallum swamps and streams from the Richmond River area of northern New South Wales and in Queensland north of Brisbane from the Caboolture Shire to Tin Can Bay including Moreton and Fraser Islands.

Habitat: Usually among emergent vegetation or near vertical or undercut banks among fine rootlets of riparian vegetation.

Threatening processes: Loss of habitat from residential housing development, exotic pine plantations, mining and agriculture. May be affected by the introduced gambusia (*Gambusia holbrooki*), especially in the Noosa River.

Red finned blue-eye, *Scaturiginichthys vermeilipinnis* Ivantsoff et al. (1991)

Current distribution: Only known from four artesian springs in the Edgebaston Spring Complex near Aramac, central Queensland.

Habitat: Shallow springs in Mitchell grass country.

Threatening processes: Interaction with introduced gambusia (*Gambusia holbrooki*). Habitat destruction due to trampling by both domestic and feral stock. Habitat destruction due to excavation of springs for stock watering and possible overcollection for aquarium trade.

Vulnerable

Saddled galaxias, *Galaxias tanycephalus* Fulton (1978)

Current distribution: Arthur's Lake and Woods Lake, central plateau, Tasmania.

Habitat: Among rocks at margins of lakes. Larval stages pelagic.

Threatening processes: Predation by brown trout (*Salmo trutta*). Possibly nutrient enrichment of Woods Lake.

Flinders Ranges gudgeon, *Mogurnda* sp.

Current distribution: Waterways of Gammons Ranges National Park in North Flinders Ranges, South Australia.

Habitat: Small isolated waterholes.

Threatening processes: Feral goat populations may have caused eutrophication of pools. Goats have been removed.

Variegated pigmy perch, *Nannoperca variegata* Kuiter and Allen (1986)

Current distribution: Only found in Ewens Ponds near Mt Gambier, South Australia and Glenelg River near Winnap, Western Australia.

Habitat: Clear flowing streams or ponds with abundant aquatic vegetation.

Threatening processes: Destruction of habitat including removal of aquatic and riparian vegetation and drainage of swamps. Interactions with introduced species including gambusia (*Gambusia holbrooki*), redfin (*Perca fluviatilis*), brown trout (*Salmo trutta*), rainbow trout (*Oncorhynchus mykiss*) and carp (*Cyprinus carpio*).

Honey blue-eye, *Pseudomugil mellis* Allen and Ivantsoff (1982)

Current distribution: Lakes and creeks on Fraser Island and on mainland between Caboolture and Tin Can Bay, south eastern Queensland.

Habitat: Among or near emergent aquatic vegetation in streams and lakes usually associated with wallum swamps.

Threatening processes: Habitat destruction by residential development, forestry operations, agriculture and mining. Interactions with introduced gambusia (*Gambusia holbrooki*) also a threat together with collection for aquarium trade.

Potentially Threatened

Dalhousie hardyhead, *Craterocephalus dalhousiensis* Glover (1989)

Current distribution: Known from 7 springs within the Dalhousie Springs Complex, South Australia.

Habitat: Medium to large springs with warm, flowing water.

Threatening processes: Potential reduction in artesian flows, pollution of surface waters, establishment of exotic fish and uncontrolled effects of tourists.

Murray hardyhead, *Craterocephalus fluviatilis* Ivantsoff et al. (1987)

Current distribution: Restricted to Kerang Lakes area, Victoria, and possibly the nearby Murray River.

Habitat: Margins of slow, lowland rivers and in lakes, billabongs and backwaters among aquatic plants and over gravel beds.

Threatening processes: Not known but general comments on threats outlined in this paper probably apply.

Glovers Hardyhead, *Craterocephalus gloveri* Crowley and Ivantsoff (1990)

Current distribution: Known from 4 springs within Dalhousie Springs Complex, South Australia.

Habitat: Medium to large springs with warm flowing water.

Threatening processes: Potential reduction in artesian flows, pollution of surface waters, establishment of exotic fish and uncontrolled effects of tourists.

Swamp galaxias, *Galaxias parvus* Frankenberg (1968)

Current distribution: Restricted to the headwaters of the Gordon and Huon Rivers in south western Tasmania.

Habitat: Open shallow stretches of rivers or vegetation margins of swamps, quiet pools and backwaters.

Threatening processes: Not known at this stage.

Dwarf galaxias, *Galaxiella pusilla* Mack (1936)

Current distribution: Limited to north eastern Tasmania, Flinders Island, Victoria and South Australia. In Victoria, main populations are in the Grampians and four sites around Melbourne. In South Australia found in swamps and drains in the south east of the State.

Habitat: Slow flowing waters of swamps and drains or backwaters of creeks, often among aquatic vegetation in shallow waters.

Threatening processes: Habitat destruction from draining of swamps, channelisation and removal of aquatic and riparian vegetation, and interactions with introduced species, particularly gambusia (*Gambusia holbrooki*).

Non-parasitic lamprey, *Mordacia praecox* Potter (1968)

Current distribution: Known only from the Moruya and Tuross Rivers in south eastern New South Wales and possibly from the La Trobe River in Victoria although this has not been confirmed.

Habitat: Not known.

Threatening processes: Details not known, general habitat degradation.

Dalhousie catfish, *Neosilurus* n.sp.

Current distribution: Known from 5 springs within the Dalhousie Springs Complex, South Australia.

Habitat: Medium to large springs with warm, flowing water.

Threatening processes: Potential reduction in artesian flows, pollution of surface waters, establishment of exotic fish and uncontrolled effects of tourists.

Yarra pigmy perch, *Edelia obscura* Klunzinger (1872)

Current distribution: A few locations in south west Victoria and south west of South Australia.

Habitat: Slow flowing stream or still water among aquatic vegetation.

Threatening processes: General destruction of habitat including removal of aquatic and riparian vegetation, and drainage of swamps. Interactions with introduced species including gambusia (*Gambusia holbrooki*), redfin (*Perca fluviatilis*), brown trout (*Salmo trutta*) and carp (*Cyprinus carpio*).

Australian grayling, *Prototroctes maraena* Gunther (1864)

Current distribution: Patchy distribution in coastal rivers from the Grose River, west of Sydney, through New South Wales, Victoria and eastern South Australia. Also occurs in Tasmania and on King Island in Bass Strait. Often found in small numbers only.

Habitat: A diadromous species with larvae apparently being swept to estuaries and returning to freshwater after 4-6 months. Freshwater habitats include large and small coastal rivers. It is a midwater species.

Threatening processes: Barriers to movement caused by dams. River regulation. Habitat degradation caused by siltation from catchment erosion, and stream channel damage caused by sand and gravel extraction. Possible predation by brown trout (*Salmo trutta*) on juveniles.

Indeterminate

Macquarie perch, *Macquaria australasica* Cuvier and Valenciennes (1830)

Current distribution: Three genetic stocks are presently recognised:

- Shoalhaven stock, east of the Great Dividing Range in New South Wales. Possibly limited to lower reaches of Shoalhaven River system.
- Hawkesbury stock, widespread in Hawkesbury River in New South Wales.
- Murray-Darling stock, known from the Murrumbidgee and Lachlan Rivers in New South Wales and from the Loddon, Goulburn, Ovens and Mitta Mitta Rivers in Victoria. May have been stocked in impoundments in Shoalhaven and Hawkesbury River. Also stocked in Wannon, Barwon and Yarra Rivers.

Habitat: A riverine, schooling species, prefers deep rocky holes with considerable cover.

Threatening processes: Destruction of habitat by siltation of spawning sites, blanketing of riffle areas (invertebrate food source) and infilling of deep holes. Interaction with introduced species including brown trout (*Salmo trutta*) and redfin (*Perca fluviatilis*). Increased nutrient loads associated with urban development, river regulation and overfishing.

Blind cave eel, *Ophisternon candidum* Mees (1962).

Current distribution: Not well known. In the last 15 years has only been observed in two locations, Gnamma Hole and Mowbowra Well in Western Australia. Probably distributed throughout the subterranean system from just south of Exmouth, north to North West Cape and south to Yardie Creek.

Habitat: Subterranean caverns and fissures.

Threatening processes: Little known but potential to degrade water quality of subterranean waters has already been demonstrated by Mowbowra Well being filled with rubbish, and siltation of Dozer Cave (connected to Gnamma Hole) due to runoff from old mining operations. Both these sites have been rehabilitated.

Reference

McDowall, R.M. and R.S. Frankenberg (1981). The Galaxiid fishes of Australia. *Records of the Australian Museum* **33(10)**, 443-605.

APPENDIX 2: WORKSHOP PROGRAM

SUSTAINABLE FISHERIES THROUGH SUSTAINING FISH HABITAT

(Convenor: Barry Bruce South Australian Department of Fisheries)

Day 1 (Wednesday, 12 August 1992)

0900 – 0915 SESSION 1—INTRODUCTION—Dr Julian Pepperell, President, ASFB

0915 – 1000 Keynote Address—Dr Stanley Moberly, Northwest Marine Technology, Seattle

"Fish Habitat is Where It's At: It is more fun to fight over more fish than less fish."

1000 – 1030 Morning Tea

1030 – 1230 SESSION 2—A MANAGER'S VIEW OF FISH HABITAT

Chairperson: Rob Lewis, Department of Fisheries, SA

Rapporteur: Gina Newton, Bureau of Rural Resources, ACT

Panel Speakers:

Karen Edyvane, Department of Fisheries, SA "An ecosystem approach to marine fisheries management"

Ross Winstanley, Marine Science Laboratories, VIC "Estuarine issues from the manager's viewpoint"

Peter Jackson, QDPI Fisheries, QLD "Freshwater protection—a manager's perspective"

1230 – 1330 Lunch

1330 – 1500 SESSION 3—ORGANISMS AND ENVIRONMENTAL RELATIONSHIPS I—Case Studies

Chairperson: Peter Gehrke, Inland Fisheries Research Station, NSW

Rapporteur: Gary Thorncraft, Fisheries Research Institute, NSW

Panel Speakers:

Ian Poiner, CSIRO Fisheries QLD "Maintain or modify – alternative views of managing critical fisheries habitat"

Stephen Swales, Fisheries Research Institute, NSW "Rehabilitation, mitigation and restoration of fish habitat in regulated rivers"

Mick Bales, Department of Water Resources, NSW "A habitat fit for fish – an aim of biomanipulation"

Rick Morton, WBM Oceanics, QLD "Enhancement of estuarine habitats in association with development"

1500 – 1530   Afternoon Tea

1530 – 1700   **SESSION 4—ORGANISMS AND ENVIRONMENTAL RELATIONSHIPS II—Key Variables and Broad Based Issues**

Chairperson: Bryan Pierce, Department of Fisheries, SA

Rapporteur: Mervi Kangas, Department of Fisheries, SA

Panel Speakers:

Peter Young, CSIRO Fisheries, TAS "Defining key factors relating marine fishes and their habitats"

Neil Loneragan, CSIRO Fisheries, QLD "Defining key factors relating fish populations in estuaries and their habitat"

John Koehn, Arthur Rylah Institute, VIC "Freshwater fish habitats : key factors and methods to determine them"

1700 – 1730   Discussion of Day 1

Rapporteur: David Smith, Marine Science Laboratories, VIC

Day 2 (Thursday 13 August 1992)

0900 – 1030   **SESSION 5—IMPACT OF HUMAN ACTIVITIES ON HABITAT AND FISHERIES**

Chairperson: Murray McDonald, Marine Science Laboratories, VIC

Rapporteur: Patrick Coutin, Marine Science Laboratories, VIC

Panel Speakers:

Robert Campbell, CSIRO Fisheries, TAS "Effects of trawling on the marine community on the North West Shelf of Australia and implications for sustainable fisheries management"

Greg Jenkins, Victorian Institute of Marine Science, VIC "Ecological basis for parallel declines in seagrass habitat and catches of commercial fish in Western Port Bay, Victoria"

Martin Mallen-Cooper, Fisheries Research Institute, NSW "Habitat changes and declines of freshwater fish in Australia : what is the evidence and do we need more?"

1030 – 1100   Morning Tea

1100 – 1230   **SESSION 6—ALTERNATIVE USES (economic, social and political)—KEY VALUES**

Chairperson: Russell Reichelt, Bureau of Rural Resources, ACT

Rapporteur: John Robertson, GBRMPA, QLD

Panel Speakers:

David Pollard, Fisheries Research Institute, NSW "Maximising the potential for both sustainable fisheries and alternative uses of fish habitat through marine harvest refugia"

Andrew Staniford, Department of Mines and Energy, SA "Alternative uses of aquatic habitats : an economic viewpoint"

1230 – 1330   Lunch

1330 – 1530   **SESSION 7—MANAGEMENT (amelioration, enhancement, conservation)**

Chairperson: Barbara Richardson, NSW Fisheries, NSW

Rapporteur: David Pollard, Fisheries Research Institute, NSW

Panel Speakers:

Bob O'Boyle, Bedford Institute of Oceanography, CANADA "The management of the marine habitat"

Jenny Burchmore, NSW Fisheries, NSW "Management of the estuarine habitat"

Wayne Fulton, Inland Fisheries Commission, TAS "Managing freshwater fish habitat"

1530 – 1600   Afternoon Tea

1600 – 1700   **GENERAL DISCUSSION**

Chairperson: Peter Young, CSIRO Fisheries, TAS

Rapporteur: Peter Last, CSIRO Fisheries, TAS

Panel Speakers:

Rob Lewis, Department of Fisheries, SA

Peter Gehrke, Inland Fisheries Research Station, NSW

Bryan Pierce, Department of Fisheries, SA

Murray MacDonald, Marine Science Laboratories, VIC

Russell Reichelt, Bureau of Rural Resources, ACT

Barbara Richardson, NSW Fisheries, NSW

1700 – 1730   **SUMMING UP**

Dr Stanley Moberly (Keynote Speaker)

APPENDIX 3: LIST OF PARTICIPANTS

ALLEN K R — 20/8 Waratah Street
CRONULLA NSW 2230

ANDERSON J R — Univ. of New England—Nthn Rivers Campus
PO Box 157
LISMORE NSW 2480

BAKER D L — Sth. Aust. Res. & Dev. Inst.—Fisheries
GPO Box 1625
ADELAIDE SA 5001

BAKER J L — Sth. Aust. Res. & Dev. Inst.—Fisheries
GPO Box 1625
ADELAIDE SA 5001

BALES M T — Dept. Water Resources
10 Valentine Ave
PARRAMATTA NSW 2150

BAULCH G — Dept. Primary Industry & Fisheries
P.O. Box 990
DARWIN NT 0801

BENSON C G — Dept. Applied Sciences, Univ. Tasmania
P.O. Box 1214
LAUNCESTON TAS 7250

BERTONI M — Bureau of Resource Sciences
P.O. Box E11, Queen Victoria Tce,
CANBERRA ACT 2600

BIRD F L — Dept. of Zoology,, Univ. of Melbourne
PARKVILLE VIC 3052

BRIDGE N — School of Biol. Sciences, Flinders Univ.
Sturt Rd,
BEDFORD PARK SA 5042

BROWN L P — Marine Sciences Laboratories
P.O. Box 114,
QUEENSCLIFF VIC 3225

BRUCE B D — Sth. Aust. Res. & Dev. Inst.-Fisheries
GPO Box 1625
ADELAIDE SA 5001

BURCHMORE J — NSW Fisheries
P.O. Box 4805
ST. LEONARDS NSW 2065

CADWALLADER P L — Dept. of Conservation & Environment
Private Bag 20
ALEXANDRA VIC 3714

CAMERON D S — Sthn Fisheries Centre Q.D.P.I.
P.O. Box 76
DECEPTION BAY QLD 4508

CAMPBELL R A — CSIRO Division of Fisheries
GPO Box 1538
HOBART TAS 7001

CLAGUE C I — 67 Old Smithfield Rd
Freshwater
CAIRNS QLD 4870

CROSS H — NSW Dept. Water Resources
P.O. Box 370
PARRAMATTA NSW 2124

CONNOLLY R — Dept. of Zoology, Univ. of Adelaide
GPO Box 598
ADELAIDE SA 5001

COUTIN P C — Marine Sciences Laboratories
P.O. Box 114,
QUEENSCLIFF VIC 3225

DAVIES C R — School of Biol. Sci., James Cook Univ.
Post Office
TOWNSVILLE QLD 4811

DIPLOCK J — NSW Fisheries
P.O. Box 4805
ST. LEONARDS NSW 2065

DIXON P — Centre for Marine Sciences
Univ. of NSW P.O. Box 1
KENSINGTON NSW 2034

EDYVANE K — Sth. Aust. Res. & Dev. Inst.-Fisheries
GPO Box 1625
ADELAIDE SA 5001

EVANS K — Sth. Aust. Res. & Dev. Inst.-Fisheries
GPO Box 1625
ADELAIDE SA 5001

FRUSHER S D — Div. of Sea Fisheries, D.P.I.F. & E.
P.O. Box 192B
HOBART TAS 7001

FULTON W	Inland Fisheries Commission 127 Davey St HOBART TAS 7000	JONES K	Sth. Aust. Res. & Dev. Inst.- Fisheries GPO Box 1625 ADELAIDE SA 5001
GEHRKE P C	NSW Fisheries P.O. Box 182 NARRANDERA NSW 2700	JORDAN A R	Dept. of Primary Industries & Fisheries Crayfish Point TAROONA TAS 7053
GLAISTER J P	Sthn Fisheries Centre Q.D.P.I. P.O. Box 76 DECEPTION BAY QLD 4508	KAILOLA P J	Bureau of Resource Sciences P.O. Box E11, Queen Victoria Tce, CANBERRA ACT 2600
GOMON M F	Museum of Victoria 285 Russell St MELBOURNE VIC 3000	KANGAS M I	Sth. Aust. Res. & Dev. Inst.- Fisheries GPO Box 1625 ADELAIDE SA 5001
HALL N G	Bernard Bowen Fisheries Research Institute P.O. Box 20 NORTH BEACH WA 6020	KENYON R A	CSIRO Division of Fisheries P.O. Box 120 CLEVELAND QLD 4163
HANCOCK D A	29 Woodlands Way QUINDALUP WA 6281	KINLOCH M A	Sth. Aust. Res. & Dev. Inst.- Fisheries GPO Box 1625 ADELAIDE SA 5001
HAYWOOD M	CSIRO Division of Fisheries P.O. Box 120 CLEVELAND QLD 4163		
HOBDAY D	Marine Sciences Laboratories P.O. Box 114, QUEENSCLIFF VIC 3225	KOEHN J D	Arthur Rylah Institute 123 Brown St. HEIDELBERG VIC 3084
HOUGHTON B R	6 Fernley Ave McLEOD VIC 3085	KUITER R	110 Kananook Ave SEAFORD VIC 3198
HUGHES D	School of Biol. Sciences, Macquarie Univ. NORTH RYDE NSW 2109	KUKOLIC K	Dept. of Environment, Land & Planning P.O. Box 1065 TUGGERANONG ACT 2901
JACKSON B	Sth. Aust. Res. & Dev. Inst.- Fisheries GPO Box 1625 ADELAIDE SA 5001	LAST P R	CSIRO Division of Fisheries GPO Box 1538 HOBART TAS 7001
JACKSON G	Sth. Aust. Res. & Dev. Inst.- Fisheries GPO Box 1625 ADELAIDE SA 5001	LAURENSON F A	Fisheries Research Institute P.O. Box 21 CRONULLA NSW 2230
JACKSON P D	Sthn Fisheries Centre Q.D.P.I. P.O. Box 76 DECEPTION BAY QLD 4508	LEADBITTER D	Ocean Watch P.O. Box 247 PYRMONT NSW 2009
JENKINS G P	P.O. Box 138 QUEENSCLIFF VIC 3225	LENANTON R C J	Bernard Bowen Fisheries Research Institute P.O. Box 20 NORTH BEACH WA 6020
JOLL C P	Australian Fisheries Management Authority P.O. Box 7051 CANBERRA CENTRE ACT 2610	LEWIS R J	R.M.B. 5462 SHEPPARTON VIC 3631

LEWIS R K	Sth. Aust. Res. & Dev. Inst.-Fisheries GPO Box 1625 ADELAIDE SA 5001	NEIRA F J	School of Biol. & Env. Sciences Murdoch Univ. MURDOCH WA 6150
LIGHT B R	P.O. Box 244 ROSANNA VIC 3084	NEWTON G M	Bureau of Resource Sciences P.O. Box E11, Queen Victoria Tce, CANBERRA ACT 2600
LONERAGAN N	CSIRO Division of Fisheries P.O. Box 120 CLEVELAND QLD 4163	NOWARA G	Australian Maritime College P.O. Box 986 LAUNCESTON TAS 7250
LOWRY M B	School of Biol. Sciences, Univ. of NSW P.O. Box 1 KENSINGTON NSW 2034	O'BOYLE R N	Dept. of Fisheries & Oceans P.O. Box 550 HALIFAX NOVA SCOTIA CANADA B3J 257
LUSCOMBE M	11 Folkstone Rd BRIGHTON SA 5048	O'CONNOR J P	Arthur Rylah Institute 123 Brown St. HEIDELBERG VIC 3084
MACDONALD C M	Marine Sciences Laboratories P.O. Box 114, QUEENSCLIFF VIC 3225	O'CONNOR W G	Arthur Rylah Institute 123 Brown St. HEIDELBERG VIC 3084
MALLEN-COOPER M	Fisheries Research Institute P.O. Box 21 CRONULLA NSW 2230	O'KANE C J	Fisheries Research Institute P.O. Box 21 CRONULLA NSW 2230
MATERIA C J	Museum of Victoria 285 Russell St MELBOURNE VIC 3000	OKS E M	Dept. of Environment & Planning GPO Box 667 ADELAIDE SA 5001
MESSNER K	P.O. Box 405 CEDUNA SA 5690	OLSEN A M	11 Orchard Grove, NEWTON SA 5074
MISKIEWICZ A G	NSW Water Board - Env. Projects Unit P.O. Box 453 SYDNEY SOUTH NSW 2000	O'MAHONY D J	Arthur Rylah Institute 123 Brown St. HEIDELBERG VIC 3084
MOBERLY S J	Northwest Marine Technology Inc., P.O. Box 99488 SEATTLE WA 98199-0488 USA	PARAS G	Dept. of Wildlife Reserves La Trobe University BUNDOORA VIC 3083
		PEPPERELL J G	Pepperell Research Pty. Ltd. P.O. Box 818 CARINGBAH NSW 2229
MORRISON A K	Marine Sciences Laboratories P.O. Box 114, QUEENSCLIFF VIC 3225	PIERCE B E	Sth. Aust. Res. & Dev. Inst.-Fisheries GPO Box 1625 ADELAIDE SA 5001
MORTON R M	WBM Oceanics Australia Pty. Ltd. P.O. Box 203 SPRING HILL QLD 4004	PITCHER C R	CSIRO Division of Fisheries P.O. Box 120 CLEVELAND QLD 4163
MURRAY A J	Dept. of Conservation & Environment P.O. Box 260 ORBOST VIC 3225	POINER I R	CSIRO Division of Fisheries P.O. Box 120 CLEVELAND QLD 4163
MUSA J	Centre for Marine Sciences Univ. of NSW P.O. Box 1 KENSINGTON NSW 2034	POLLARD D A	Fisheries Research Institute P.O. Box 21 CRONULLA NSW 2230

POOLE R	Museum of Victoria 528 Swanston St. MELBOURNE VIC 3000	STEWART B D	Zoology Dept University of Melbourne PARKVILLE VIC 3052
PRINCE J D	P.O. Box 209 LEEDERVILLE WA 6007	STOBUTSKI I C	24 Wilmett St TOWNSVILLE QLD 4810
PUCKRIDGE J T	Dept. of Zoology, Univ. of Adelaide GPO Box 598 ADELAIDE SA 5001	SUCKLING G C	Dept. of Conservation & Environment 5/240 Victoria Pde MELBOURNE VIC 3002
PULLEN G	Marine Laboratories Div. of Sea Fisheries TAROONA TAS 7053	SUTTON C A	Centre for Marine Sciences Univ. of NSW P.O. Box 1 KENSINGTON NSW 2034
REICHELT R	Bureau of Resource Sciences P.O. Box E11, Queen Victoria Tce, CANBERRA ACT 2600	SWALES S	Fisheries Research Institute P.O. Box 21 CRONULLA NSW 2230
RICHARDSON B A	NSW Fisheries P.O. Box 4805 ST. LEONARDS NSW 2065	TAY H C	Centre for Marine Sciences Univ. of NSW P.O. Box 1 KENSINGTON NSW 2034
ROBERTSON J	Great Barrier Reef Marine Park Authority P.O. Box 1379, TOWNSVILLE. QLD 4810	THORNCRAFT G A	Fisheries Research Institute P.O. Box 21 CRONULLA NSW 2230
ROWLING K R	Fisheries Research Institute P.O. Box 21 CRONULLA NSW 2230	TRENDALL J	West Beach Aquaculture Pty. Ltd. P.O. Box 559 GLENELG SA 5045
RUSSELL D J	Nthn Fisheries Centre. Q.D.P.I. P.O. Box 5396 CAIRNS QLD 4870	UNMACK P	161 High St DONCASTER VIC 3108
		VANCE D J	CSIRO Division of Fisheries P.O. Box 120 CLEVELAND QLD 4163
SADDLIER S R	Arthur Rylah Institute 123 Brown St. HEIDELBERG VIC 3084	WALKER T I	Marine Sciences Laboratories P.O. Box 114, QUEENSCLIFF VIC 3225
SHEPHERD M A	School of Biol. Sciences, Macquarie Univ. NORTH RYDE NSW 2109	WHITE G F	Dept. Primary Industry & Fisheries P.O. Box 990 DARWIN NT 0801
SHEPHERD S A	Sth. Aust. Res. & Dev. Inst.-Fisheries GPO Box 1625 ADELAIDE SA 5001	WINSTANLEY R H	Marine Sciences Laboratories P.O. Box 114, QUEENSCLIFF VIC 3225
SMITH D C	Marine Sciences Laboratories P.O. Box 114, QUEENSCLIFF VIC 3225	YOUNG P C	CSIRO Division of Fisheries GPO Box 1538 HOBART TAS 7001
STANIFORD A J	Office of Energy Planning 30 Wakefield St ADELAIDE SA 5000		
STAPLES D	Bureau of Resource Sciences P.O. Box E11, Queen Victoria Tce, CANBERRA ACT 2600		